WIND+ SUN+ NUCLEAR

A positive vision for Australia's energy future

DAVID WILSON

Published in Australia by Sid Harta Books & Print Pty Ltd,

ABN: 34632585293

23 Stirling Crescent, Glen Waverley, Victoria 3150 Australia

Telephone: +61 3 9560 9920, Facsimile: +61 3 9545 1742

E-mail: author@sidharta.com.au

First published in Australia 2025

This edition published 2025

Cover design, typesetting: WorkingType (www.workingtype.com.au)

ISBN: 978-1-923439-90-0

About the Author

David Wilson, born in Dalby, Queensland, in 1947, has had a lifelong passion for tackling complex challenges with pragmatic solutions. As a career civil engineer with decades of experience in the oil and gas industry, David's professional journey spanned the globe, from offshore engineering projects in Singapore to leading his own firm in London. His unique perspective is shaped by a deep understanding of energy systems, technical innovation and Australia's evolving climate and energy needs.

In *Wind + Sun + Nuclear*, David applies his engineering expertise and analytical rigour to one of the country's most pressing issues: the future of energy. He critiques Australia's heavy reliance on renewables, highlighting their potential for unreliability without proper baseload solutions, and makes a compelling case for the inclusion of nuclear power in the nation's energy mix.

Now retired on the Sunshine Coast, David remains an active voice in the climate and energy conversation, advocating for practical, science-based strategies to ensure a sustainable and reliable energy future for Australia.

Wind + Sun + Nuclear is David's second book. He has also written a memoir of his life in the oil and gas industry, *Never Fear the Spills*, also published by Sid Harta.

Other titles by David Wilson

Disclaimer

All best endeavours have been undertaken to secure permission to use the reproduced material in this book.

Abbreviations and Symbols

Term	Definition
a	albedo
AAAS	American Association for the Advancement of Science
AEMO	Australian Energy Market Operator
AIMS	Australian Institute of Marine Science
A_p	the projected surface area of Earth, in m^2
AR6	Assessment Report 6
A_s	the surface area of an object, in m^2
Bn	billion
$CaCO_3$	calcium carbonate
CaO	calcium oxide
$Ca(OH)_2$	calcium hydroxide
capex	capital expenditure
CCA	Climate Change Authority
CFC	chlorofluorocarbon
CH_4	methane
CO_2	carbon dioxide
CO_3^{2-}	carbonate

Term	**Definition**
CMIP6	Coupled Model Intercomparison Project Phase 6
CNNC	China National Nuclear Corporation
COP21	Conference of the Parties UNFCCC Paris 2015
COP26	Conference of the Parties UNFCCC Glasgow 2021
CPRS	Carbon Pollution Reduction Scheme
°C	degrees Celsius
DCCEEW	Department of Climate Change, Energy, the Environment and Water
E_{in}	the energy Earth receives from the Sun as solar radiation
ENS	École Normale Supérieure
E_{out}	the energy Earth transmits into outer space as infrared radiation
EPBC Act	*Environmental Protection and Biodiversity Conservation Act*
FAR	First Assessment Report
FEED	front-end engineering and design
GBR	Great Barrier Reef
g/mol	grams per mole
GHG	greenhouse gas
Gt	gigatonne, billion tonnes
GW	gigawatt
H^+	hydrogen ion
H_2CO_3	carbonic acid
H_2O	water
HCO_3^-	bicarbonate

Term	Definition
IPCC	Intergovernmental Panel on Climate Change
JCU	James Cook University
K	degrees kelvin
K_2CO_3	potassium carbonate
K_2SO_4	potassium sulphate
kW	kilowatt
kWh	kilowatt hour
LCOE	levelised cost of electricity
$MgCO_3$	magnesium carbonate (magnesium alba)
MgO	magnesium oxide (magnesia)
$MgSO_4$	magnesium sulphate
MW	megawatt
MWh	megawatt hour
Mt	megatonne, million tonnes
N_2O	nitrous oxide, laughing gas
NDC	Nationally Determined Contribution
NEG	National Energy Guarantee
OCGT	open-cycle gas turbine
PhD	Doctor of Philosophy
ppm	parts per million
REZ	Renewable Energy Zone
S	the solar constant (approximately 1367 W/m^2)
SMR	small modular reactor (nuclear)
SO_2	sulphur dioxide
T	the absolute temperature of an object, in degrees kelvin (K)

Term	**Definition**
T_e	Earth's equilibrium temperature. The temperature (T_e) at which the energy Earth receives from the Sun as solar radiation (E_{in}) equals the energy Earth transmits into outer space as infrared radiation (E_{out}).
TWh	terawatt hour
Tt	teratonne, trillion tonnes
TVA	Tennessee Valley Authority
UN	United Nations
UNEP	United Nations Environment Programme
UNFCCC	United Nations Framework Convention on Climate Change
USA	United States of America
UV	ultraviolet
W/m^2	watts per square metre
WMO	World Meteorological Organization
ΔT	temperature change
€	Euro
σ	the Stefan–Boltzmann constant, approximately 5.67×10^{-8} $W/(m^2 \cdot K^4)$

Contents

Chapter 1

Genesis

In the beginning God created the heaven and the earth. And the earth was without form, and void; and darkness was upon the face of the deep. And the Spirit of God moved upon the face of the waters.[1]

It may have happened this way; however, since these apocryphal words were written over two thousand years ago, science has flourished and fostered new branches, and our understanding of the process of Earth's formation has advanced.

We now know our solar system formed from a giant molecular cloud of dust and gas, known as the solar nebula, 4.6 billion years ago. Gravitational forces caused the nebula to collapse, forming a rotating disc with the Sun at its centre. Within the disc, small particles collided and accreted and the planets with their moons were gradually formed.

For the first half billion years Earth differentiated into layers

1 Genesis 1:1-2, Miles Coverdale Bible, 1535 Edition.

based on density, with heavier materials sinking towards the centre and lighter materials rising to the surface, thereby developing a core, mantle, crust structure. Coincidentally there was an intense asteroid and comet bombardment, which not only reshaped Earth's surface but most likely endowed it with the essentials of life – water and organic molecules.

Water was also a product of volcanic activity and in the next half billion years water vapour released from volcanic activity condensed to form Earth's oceans. The volcanoes were also responsible for releasing carbon dioxide (CO_2), methane and ammonia, thereby forming a primordial atmosphere. We had probably arrived at the point where life originated in hydrothermal vents, deep-sea vents or warm shallow pools.

At first, simple organic molecules gradually assembled into more complex structures, leading to the emergence of the first single-celled organisms. Since then, life on Earth has diversified and evolved into a wide range of organisms ranging from simple bacteria to complex multicellular organisms. During this time Earth's surface and atmosphere have undergone significant changes due to geological processes, biological activity, climate fluctuations and external influences such as asteroid impacts.

As humans, we argue that the crowning achievement of this genesis was the emergence of *Homo sapiens* in Africa some 300 000 years ago. Then, around 70 000 years ago, driven by various factors, including climate change, population growth, the search for new resources and sheer curiosity, they gradually spread from Africa to other parts of the world.

Early human societies were predominantly hunter-gatherer

communities, relying on hunting wild animals and gathering wild plants for sustenance. However, around 10 000 years ago humans began cultivating plants and raising animals. This led to settled agricultural communities and the production of surplus food, population growth, specialisation of labour, the development of more complex social structures and the pursuit of learning.

Early civilisations like the Egyptians, Babylonians and Greeks laid the groundwork for mathematical concepts such as arithmetic, geometry and algebra. Early astronomical observations enabled the Babylonians to make significant contributions to understanding celestial movements, leading to the creation of the first calendars. Various forms of traditional medicine came soon after, often blending empirical observations with spiritual beliefs. Greek philosophers like Aristotle laid down the foundations of Western philosophy, which included inquiry into nature, metaphysics, ethics and logic. Euclid's *Elements*[2] became the foundational work in geometry, while Archimedes made significant contributions to mathematics and physics. Greek astronomers such as Ptolemy developed geocentric models of the universe, which dominated astronomical thought for centuries.

The Renaissance (1400–1650) marked a period of renewed interest in scientific inquiry, with figures like Leonardo da Vinci and Galileo Galilei making significant contributions to observational science. Nicolaus Copernicus proposed a heliocentric model of the solar system, challenging the

---

2 Euclid, *The Elements*, translated by Sir Thomas L. Heath, 3 volumes, Cambridge University Press, Cambridge, 1908.

geocentric view and laying the groundwork for modern astronomy. Figures like Isaac Newton, Johannes Kepler and Galileo Galilei made groundbreaking discoveries in physics, astronomy and mathematics, leading to the development of the scientific method and the birth of modern science. This included climatology.

Climatology emerged as a distinct scientific discipline during the late nineteenth and early twentieth centuries. While humans have been observing and recording weather patterns for centuries, the systematic study of climate and its underlying processes began to take shape with the development of more sophisticated observational techniques, advances in mathematics and physics, and the establishment of meteorological institutions such as the Met Office Hadley Centre in England. Its Central England Temperature dataset (Figure 1.1) is the world's longest continuous instrumental temperature record, gathering data from various sources in the Midlands of England since 1659. This includes data from the world's longest-running weather station atop the Radcliffe Observatory at the University of Oxford. Weather observations, including temperature, pressure, wind direction and precipitation, have been recorded at this station continuously ever since Thomas Hornsby, mathematician and astronomer, established it in 1772.

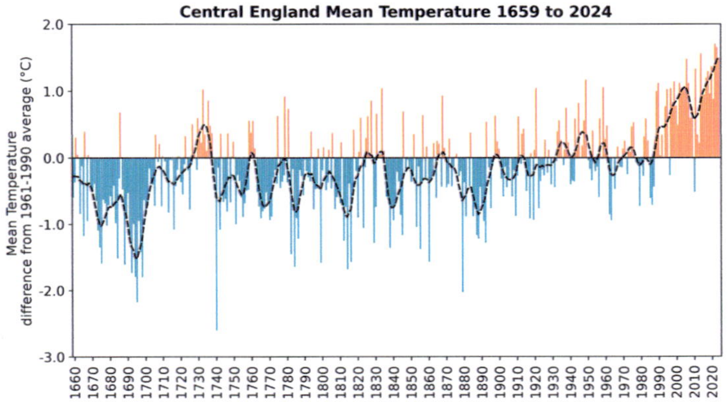

Figure 1.1 Central England mean temperature 1659 to 2023, expressed as deviation from the 1961–1990 average[3]

Three years later Joseph Black discovered carbon dioxide (then known as 'fixed air'), and the story of climate change began. Atmospheric CO_2 concentration was a steady 285 parts per million (ppm), equating to a total of 2.23 trillion tonnes (Tt) of CO_2 in the atmosphere; it had been at this level for at least 10 000 years. At the time the global population of less than a billion along with natural phenomena were producing less than 40 million tonnes (Mt) of CO_2 a year.

It would be another fifty years before Joseph Fourier published his memoir on the temperature of Earth and planetary spaces in which he proposed that Earth's atmosphere acts like an insulator, trapping heat and contributing to the planet's overall temperature balance. He was right. But nothing had changed apart from the global population, which had grown to around 1.1 billion.

---

3 Central England Mean Temperature 1659 to 2023, Met Office Hadley Centre for Climate Change, UK, 2024, https://hadleyserver.metoffice.gov.uk/hadcet/cet_info_mean.html, accessed December 2024.

The first assertion that an excess of CO_2 would upset the temperature balance was made in 1856 when Eunice Newton Foote, an amateur scientist from upstate New York, suggested:

> An atmosphere of that gas [carbon dioxide] would give to our earth a high temperature; and if as some suppose, at one period of its history the air had mixed with it a larger proportion than at present, an increased temperature from its own action as well as from increased weight must have necessarily resulted.[4]

In effect, CO_2 and other greenhouse gases (GHGs) would trap heat transmitted from Earth, upset its energy equilibrium and initiate global warming.

Eunice Newton Foote's predictions were soon confirmed by European scientists, including Svante Arrhenius, who in 1896 developed a theoretical model linking increased CO_2 levels from fossil fuel combustion to potential global warming. He calculated that doubling the atmospheric concentration of CO_2 would result in an increase in Earth's average surface temperature of approximately 5–6°C (now estimated to be about 2–4.5°C). Arrhenius would not have been able to accurately measure CO_2 concentrations using the methods available at the time, but we now know it had grown to 295 ppm, equating to a total of 2.31 Tt of CO_2. The global population was 1.6 billion and the industrial revolution was well under way, pumping 1.5 Gt of CO_2 into the atmosphere every year.

---

4 Foote, Eunice, Circumstances Affecting the Heat of the Sun's Rays, *The American Journal of Science and Arts*, 1856, vol. 22, no. 66, pp 382–383.

But nobody worried. If anything, climatologists were more concerned with the prospect of cooling. By 1900, substantial progress had been made in understanding glaciation, including dating past ice ages, mapping the extent of glacial coverage and beginning to explore the causes of glacial cycles, including attempts to link Earth's orbital variations and CO_2 levels with ice ages.

Milutin Milanković, Serbian mathematician, astronomer and geophysicist, began his work on orbital variations as a prisoner on parole in Budapest in 1915. His theory of orbital variations influencing Earth's climate was eventually developed and published over several decades culminating with his most comprehensive and influential work, *Canon of Insolation and the Ice-Age Problem*, published in 1941.[5] The book presented calculations describing how variations in Earth's orbit and axial tilt caused by interactions with other planets in the solar system affect solar radiation received by Earth. These variations are now known as Milanković cycles and influence long-term climate patterns leading to glacial and interglacial cycles. In 1941, atmospheric CO_2 concentration was 311 ppm, and the global population of 2.3 billion was pumping 5 Gt of it into the atmosphere every year.

Thirty years later an analysis of deep ocean sediment cores confirmed Milanković's theories. These studies also showed how Milanković cycles drive long-term climate changes through their influence on atmospheric CO_2 levels and other components of the global carbon cycle. Perhaps the leader in this field was Charles David Keeling (1928–2005),

5 Milanković, Milutin, *Canon of Insolation and the Ice-Age Problem*, Gauthier-Villars, Belgrade, 1941.

best known for the 'Keeling Curve', a graph that shows the ongoing change in concentration of carbon dioxide in Earth's atmosphere since the 1950s.

The initial reading was 315 ppm. By 1970 it was 326 ppm, although there was no corresponding rise in global temperatures. The explanation for this emerged in the wake of Richard Nixon's *Clean Air Act Amendments* of 1970. This legislation established the National Ambient Air Quality Standards for major air pollutants, including sulphur dioxide (SO_2). Atmospheric SO_2 is a coolant. It is also a pollutant often leading to smog and haze over industrial areas. As the new regulations took hold and atmospheric SO_2 concentrations plummeted, blue skies returned. It thus became clear that SO_2 cooled the atmosphere more than CO_2 warmed it. By 1988, with CO_2 at 351 ppm and SO_2 virtually gone, Earth's average temperature had risen by 0.28°C. This was the first hard evidence that mankind was causing global warming.

The civilisation that had produced Shakespeare, Mozart and Einstein was fouling its own nest. It wasn't the first time. The first tangible sign mankind had the potential to damage the atmosphere emerged in the 1970s when scientists discovered chlorofluorocarbons (CFCs) released to the atmosphere were depleting the ozone layer protecting Earth from the harmful effects of ultraviolet (UV) radiation. In response to growing concerns, the international community took action, and in 1987 the Montreal Protocol on Substances that Deplete the Ozone Layer was adopted. This landmark agreement aimed to phase out the production and use of ozone-depleting substances, including CFCs, halons and other related compounds. It worked. Something similar had to be done about CO_2.

In 1988, the United Nations Environment Programme (UNEP) and the World Meteorological Organization (WMO) established the Intergovernmental Panel on Climate Change (IPCC). It brings together scientists and experts from around the world to assess and synthesise the latest scientific research on climate change. Its assessments are based on peer-reviewed literature and undergo rigorous review and scrutiny by governments and experts.

The IPCC's First Assessment Report (FAR), published in 1990, included computer climate models' assessments of the state of scientific understanding of climate change and its potential impacts. Since then, subsequent IPCC assessment reports have increasingly relied on a diverse array of climate models to project future climate scenarios under different emissions scenarios and to assess the uncertainties associated with these projections.

The Third Assessment Report, issued in 2001, included the famous 'hockey stick' diagram (Figure 1.2) illustrating the differences between various models predicting rises in Earth's surface temperature by 2100 of between 1.5 and 5.5°C above 1990 levels. These predictions shocked many at the time, but there were just as many, including the author, who were sceptical. There were also the oil and gas companies, which funded campaigns to debunk the science.

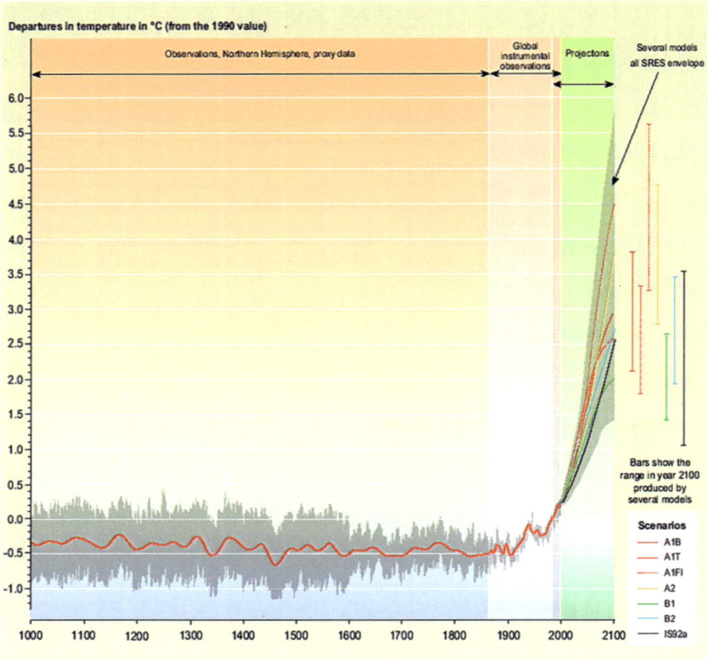

Figure 1.2 Observed global temperature data and projections for their departure from the 1990 values (in °C)[6]

Cometh the moment, cometh the man. We will never know what Al Gore would have done about global warming had he won the US presidential election in 2000, but since he didn't, he was able to throw himself into a campaign that has successfully alerted the world to a threat on par with Jules Verne's Martians. His film, *An Inconvenient Truth,*[7] released in 2006, convinced tens of millions of people worldwide that global warming was real, and a year later he shared the Nobel

6 Departure in temperature in °C (from the 1990 value), Climate Change 2001 Synthesis Report, Figure SPM-10b, p 34, The Intergovernmental Panel on Climate Change (IPCC), 2001.

7 Guggenheim, Davis (Director), *An Inconvenient Truth* (Documentary), Paramount Classics, Participant Productions, 2006.

Peace Prize with the IPCC for educating people worldwide about man-made climate change and laying the foundations for the measures needed to reverse it.

(image origin unknown)

As an Australian one might have been forgiven for thinking Gore and the IPCC had won their Nobel Prize under false pretences because there were dozens of so-called well-educated Australian politicians, including at least one Rhodes Scholar, who believed the science of climate change was baloney. More importantly, *An Inconvenient Truth* raised the inconvenient question, 'Where will we be without the fossil fuel exports that underpin our economy?' Suddenly, Australia, a stable democracy with a strong economy, world-class education system, universal healthcare, a well-developed social welfare system and a culturally tolerant society within secure borders, was facing an existential threat. A threat that has divided us ever since. At first it was denial of the science. Now it's a battle over the solution.

When the Australian Government under Liberal Prime Minister Malcolm Turnbull ratified the Paris Agreement in November 2016, it undertook to reduce Australian greenhouse gas emissions by 26–28% below 2005 levels by 2030. This fell well short of the treaty targets, not to mention international expectations, so in the lead up to the 26th Conference of the Parties to the United Nations Framework Convention on Climate Change (COP26) in 2021, Mr Turnbull's successor, Scott Morrison, committed Australia to net zero[8] emissions by 2050.

It wasn't enough. The Morrison government was defeated in the 2022 federal election, and although climate change was not the major issue, many believed Morrison's plan to achieve net zero with a grab bag of technology-led solutions was no more than a fig leaf.

With Labor it would be different. The new government moved quickly to formalise Australia's emission reduction commitments with the *Climate Change Bill 2022* which passed through Parliament in September. The Climate Change Authority (CCA), an independent statutory body, was tasked with providing analysis and recommendations on the pathways to meet Australia's emission reduction targets. Their findings were adopted by the Department of Climate Change, Energy, the Environment and Water (DCCEEW), which modelled the economic and environmental impacts of different pathways to achieve net zero by 2050. They

8 Net zero implies achieving a balance between the amount of greenhouse gases emitted into the atmosphere and the amount removed from it. With no net increase in atmospheric greenhouse gases there shouldn't be any further global warming.

concluded 'the lowest cost way to supply electricity to homes and businesses throughout Australia's transition to a net zero economy' is 'renewable energy connected by transmission, firmed with storage and backed up by gas-fired generation'.[9]

This would have been fine if nuclear options hadn't been ignored. Ignored because the Australian Labor Party has historically opposed the development of nuclear power, and there is current legislation preventing its use. Neither of these are compelling arguments and, regardless, the government is forging ahead with its plan involving the installation of 213 GW of wind and solar generation, 22 GW of flexible gas and hydro generation and 10 000 km of transmission lines ('Rewiring the Nation') by 2050.

The rollout of wind and solar has been impressive, and by mid-2024, 48 GW of the 213 GW of wind and solar to be installed on the Australian east coast by 2050 had been commissioned. Decommissioning of coal-fired power is also gathering pace. Of the 30 GW of capacity available in 2010 only 21 GW remain, and these will all be closed by 2037. But with 21 GW of coal backing up 48 GW of renewables the system is already under strain. It is difficult to imagine this will improve by 2050 with just 22 GW of flexible gas and hydro backing up a massive 213 GW of wind and solar. This may be why the opposition leader, Peter Dutton, has proposed reinforcing the network by installing nuclear reactors on seven abandoned

9 Australian Energy Market Operator (AEMO), 2024 Integrated System Plan for the National Electricity Market, A Roadmap for the Energy Transition, June 2024, https://aemo.com.au/energy-systems/major-publications/integrated-system-plan-isp/2024-integrated-system-plan-isp, accessed November 2024

coal-fired power station sites. This proposal has ignited a fierce debate, the latest in Australia's climate wars, and will feature prominently during the next election due before July 2025.

Having consumed five prime ministers, it's fair to say that climate change is one of the biggest issues in Australian politics and over the next twenty years will devour around five percent per annum of total government expenditure. To inform the debate, the history of climate science leading to the legally binding Paris Agreement of 2015 is reviewed in the following chapters. Furthermore, a climate model has been prepared showing that having achieved net zero emissions the problem of ocean acidification will continue, and this is discussed in the context of the Great Barrier Reef. Finally, Australia's response to the Paris Agreement is discussed together with a compelling argument for a role for nuclear technology, Australia's next climate divide.

Chapter 2

Joseph Black (1728–1799)

No one would have believed in the early years of the eighteenth century that mankind was under threat from a gas it was breathing daily. Certainly not a twelve-year-old leaving his family home in Bordeaux to study Latin and Greek in Belfast. But in the years to come Joseph Black took chemistry out of the kitchen, turned it into a science and discovered carbon dioxide.

Black was born on 16 April 1728. His parents were Scottish and his grandfather, Robert Gordon, had established a thriving wine export business in Bordeaux.[10] In 1726, having bought two estates and associated titles, Robert returned to Scotland to retire a grandee. He left his son-in-law in charge. John Black had done well. He and his wife Margaret and twelve of their thirteen children lived in the posh district of Chartrons with a view over the Garonne to their farmhouse and vineyards in Lormont. His eldest son, Robert, had moved to the Isle of Man where, as a partner in a trading house, he provided his father

10 Ramsay, Sir William. *The Life and Letters of Joseph Black, M.D.*, Constable and Company Ltd, London, 1918.

with a lucrative way of avoiding trade embargoes between France and Ireland.[11] There is no account of young Joseph's voyage from France, but since he had a brother in the Isle of Man it's more than likely he travelled to Belfast via Douglas in a ship carrying his father's contraband. This would have been a vessel somewhere between a Dutch fluyt of 250 British tons, such as the pilgrim's *Mayflower*, and an English bark of 300 tons, such as Captain Cook's *Endeavour*.

The voyage must have gone well since we know young Joseph spent the next four years at school in Belfast studying Latin and Greek in a private school. That must have gone well too since in 1744 he enrolled at the University of Glasgow where he spent four years completing the customary arts degree in languages and philosophy. Then, sensitive to his father's desire he take a profession, he enrolled in the school of medicine under the tutelage of Professor William Cullen.

Cullen was a grammar school boy who also studied arts at the University of Glasgow before taking up an apprenticeship with a surgeon apothecary. In early eighteenth-century Scotland these were practitioners who combined the roles of surgeon and apothecary. Their duties and responsibilities were diverse and encompassed various aspects of healthcare, though the specific roles and qualifications could vary depending on the time and location. They would diagnose and treat a wide range of illnesses and health conditions, relying on their knowledge of both surgery and pharmacology.

---

11 Irish Sea and Coastal Trade, record maintained by the Ulster Historical Foundation, Newtownards, https://ulsterhistoricalfoundation.com/irish-parliament/background-to-statutes/smuggling-quarantine-and-wrecking, accessed December 2024.

Surgical procedures of the day included amputations, wound treatment, abscess drainage, bloodletting, setting bones, fracture management and other minor interventions. Physicians would also be involved in midwifery, assisting with childbirth and providing prenatal and postnatal care. As apothecaries, they were responsible for compounding and dispensing various medicines, including herbal remedies and other pharmaceuticals. They needed to have a good understanding of pharmacology and be skilled in preparing and prescribing medicines. As we will learn later, the science of chemistry emerged from the practice of medicine.

In 1729, aged nineteen and at the end of his apprenticeship, Cullen joined a West Indies-bound trader as surgeon, no doubt honing his skills on real live patients. Then, after a year on the high seas, he jumped ship to take up a post as an assistant apothecary in London. For one interested in medicine it was the place to be. Medical advances in London in the 1730s were characterised by improvements in surgical techniques, the practice of inoculation, the establishment of hospitals and medical training institutions, and a growing interest in pharmaceutical chemistry. These efforts laid the groundwork for future medical breakthroughs, particularly in surgery and infectious disease prevention.

In 1734 a small inheritance enabled Cullen to return to Scotland and enrol in the faculty of medicine at the University of Edinburgh. This school was associated with a public hospital and at the time had a reputation as one of the world's leading medical centres, with prominent practitioners, innovative teaching methods and a focus on anatomical studies.

In 1738, having completed his studies, Cullen returned to Glasgow to enter general practice and to take his medical studies to a higher level. He was soon giving lectures in medicine and chemistry at the University of Glasgow and in 1751 was appointed chair of medicine. His work covered a broad range of medical topics, but he is particularly well known for his contributions to chemistry and his role in integrating it into medical education.

Hitherto, chemistry had been considered a subfield of natural philosophy, closely linked to alchemy. Its holy grail was not too far from the transmutation of base metals into gold and the search for the philosopher's stone. Cullen, however, in his inaugural lecture, expressed a more modern and systematic perspective on chemistry. He argued that chemistry should be studied independently of alchemy and as a separate scientific discipline with its own principles and methodologies. He believed that chemistry should focus on the understanding of the composition and properties of matter, as well as the study of chemical reactions. His approach emphasised empirical observation, experimentation and the development of clear and systematic chemical theories. While he did not fully dissociate chemistry from alchemy at this point in his career, his efforts marked an important step in the direction of modern chemistry, setting the stage for Black and later chemists and the eventual separation of chemistry from alchemy.

Black took a great interest in this approach, and his enthusiasm did not go unnoticed by Cullen who invited Black to assist him with the demonstration of chemical experiments during his lectures. Black accepted and over the next three years acquired not only a solid grounding in the 'science' of chemistry but in the art of chemical experimentation as well. The winemaker's son hadn't fallen far from the tree.

After three years study with William Cullen, Black moved to Edinburgh to complete his medical education. A common complaint of the day was bladder calculus, now known as kidney stone. Mrs Joanna Stephens, an English snake oil peddler, struck gold when Sir Robert Walpole, the prime minister, imagined her secret remedy had relieved his kidney stones. In a hypochondriacal fervour he considered the treatment should be available to all and persuaded the government to pay her five thousand pounds (present value around half a million pounds) to publish her formula. She did so in the *London Gazette* on 19 June 1739.

> My medicines are a Powder, a Decoction, and Pills. The powder consists of Egg-shells and Snails, both calcined. The decoction is made by boiling some Herbs (together with a Ball, which consists of Soap, Swines'-Cresses, burnt to a Blackness, and Honey) in water. The Pills consist of Snails calcined, Wild Carrot seeds, Burdock seeds, Ashen Keys, Hips and Hawes, all burnt to a Blackness, Soap and Honey.

But hadn't there been something like this before?

Eye of newt, and toe of frog,
Wool of bat, and tongue of dog,
Adder's fork, and blind-worm's sting,

Lizard's leg, and owlet's wing,—
For a charm of powerful trouble,
Like a hell-broth boil and bubble.
Double, double toil and trouble;
Fire burn, and caldron bubble.[12]

There had, and the Scots had moved on. Their medical fraternity was not persuaded of the efficacy of such a 'quaint and caustic' remedy, and the search for a milder alternative was initiated in Edinburgh. Robert Whytt and Charles Alston's research provided some evidence for the theory that limewater, an aqueous solution of calcium hydroxide, $Ca(OH)_2$,[13] might dissolve kidney stones. William Cullen was also interested in this theory, so when Joseph Black arrived in Edinburgh in 1751, he did so with the intention of maintaining his interest in chemistry and studying the nature of limewater for his doctoral dissertation.

Limewater is prepared by crushing high-quality limestone to a powder and mixing with one part of water. $CaCO_3$ (limestone) is the active ingredient in the over-the-counter antacid Quick-Eze, and below is the reaction we observe when adding the powder to water. The mixture effervesces as the CO_2 is released and lime crystals are precipitated. The remaining liquid is limewater.

$$CaCO_3 + H_2O \rightarrow Ca(OH)_2 + CO_2$$

$Ca(OH)_2$ was known as slaked lime in Black's time and used

12 Shakespeare, William, *MacBeth*, Act 1, 1605.
13 Actual chemical formula not known at the time.

for making mortar and plaster. When heated it undergoes thermal decomposition, water is driven off and all that remains is calcium oxide, CaO, known as 'quicklime'.

$$Ca(OH)_2 + heat \rightarrow CaO + H_2O$$

Having established a basic laboratory in the Royal Infirmary, the medical school's teaching hospital, Black's initial objective was to test the idea that fire imparted causticity to limestone, thereby turning it to quicklime, and to determine why quicklime lost its causticity on exposure to air. The eureka moment came when he floated a dish filled with quicklime in water covered by an inverted glass dome.[14] He expected the dome to capture the 'causticity' lost by the quicklime, thereby lowering the water level in the dome. Instead, the exact opposite was observed. The water level in the dome was slightly raised (most likely very 'slightly' since at the time atmospheric CO_2 concentration was just 285 parts per million). Black didn't realise it but the chemical reaction he had witnessed had extracted carbon dioxide from the air inside the dome as it turned the caustic quicklime to inert chalk (limestone). These days this reaction is simply represented as follows.

$$CaO + CO_2 \rightarrow CaCO_3$$

Of course, Black couldn't have known this but sensed he had stumbled across something and set about solving the puzzle.

14 Experiment mentioned in Ramsay, Sir William, *The Life and Letters of Joseph Black, M.D.*, Constable and Company Ltd, London, 1918, p 23.

There was, however, a slight complication. Whytt and Alston's work on limewater had reached the reporting stage, so Black decided to study $MgCO_3$, a similar medication popular in Rome. It was known as magnesium alba since the raw material was mined near Magnesia in Thessaly. Black was able to precipitate pure magnesium alba crystals in his own laboratory by combining Epsom salts and pearl ashes.

$$MgSO_4 + K_2CO_3 \rightarrow MgCO_3 + K_2SO_4$$

Having done so, he undertook a series of experiments to demonstrate that calcining magnesium alba reduced it to magnesia (magnesium oxide) by releasing a gas he referred to as 'fixed air'. He then went on to show that the magnesia thus produced, when exposed to the atmosphere, reverted to magnesium alba. He also demonstrated he could capture 'fixed air' in a flask and extinguish a candle with it. The reactions are now described as follows.

$$MgCO_3 \text{ (magnesium alba)} + \text{heat} \rightarrow MgO \text{ (magnesia)} + CO_2 \text{ (fixed air)}$$

$$MgO \text{ (magnesia)} + CO_2 \text{ (fixed air)} \rightarrow MgCO_3 \text{ (magnesium alba)}$$

Black pioneered quantitative chemical experimentation and achieved results with primitive instruments, weights and measures that have since been surpassed with modern equipment.

Apart from the mortar and pestle and the glassware that are much the same now as they were three hundred years ago, the tools of the trade were rudimentary at best. Filter paper was used to separate solids from liquids. Ceramic or metallic crucibles were used to heat substances to high temperatures and tongs were used to handle them. Blowpipes and bellows were used to direct a stream of air to control the intensity of a flame when heating substances or assisting in combustion experiments. When necessary, piston and cylinder air pumps were used to create vacuums.

Precise measurement of quantities by weight was crucial and possible with balances mostly of the beam and pan design with weights ranging through ounces, drams, scruples and grains – a grain being 65 milligrams. Other instruments available at the time were thermometers, barometers, a first-generation microscope (possibly 100X) and the human senses of smell, taste and feel. Later in life Black often coughed up blood, perhaps due to an excess intake of spirit of hartshorn (ammonia solution). But that didn't stop him keeping detailed records, and it was this attention to detail that led him to his famous discovery.

In *Dissertatio medica inauguralis, de humore acido a cibis orto, et magnesia alba,*[15] his doctoral dissertation, he described the first part of the magnesium alba experiment outlined above. Having been heated thoroughly for one hour, an ounce (28.35 grams) of desiccated magnesium alba yielded three drams and one scruple (12.96 grams) of magnesia, a loss of 54.3%. We now

15 Black, Joseph, *Humore Acido a Cibo Orto et Magnesia Alba*, (The acid humor arising from food and magnesium alba), University of Edinburgh, 1754.

know that the molecular masses of $MgCO_3$ and CO_2 are 84.32 and 44.1 g/mol, respectively, so the exact answer should have been a loss of 52.3% by weight.

Black presented his doctoral dissertation in Edinburgh on 11 June 1754. He was twenty-six years old. One hundred and fifty years later, his biographer, Sir William Ramsay, a Nobel laureate in chemistry, suggested it 'gained for him not merely the degree of Doctor of Medicine, but also brought his name before every 'philosopher' in Europe and America as that of a man who had made a discovery of more fundamental influence on the progress of chemistry than any which had previously been described'.[16]

Following his graduation Black worked as a physician until 1756 when he returned to Glasgow as professor of anatomy and lecturer in chemistry. In the latter role, he discovered the concept of latent heat. He observed that when a substance undergoes a change in state, such as from a solid to a liquid or from a liquid to a gas, heat is absorbed or released without a corresponding change in temperature. Coincidentally, James Watt, who was working as an instrument maker at the university, became interested in steam engines. Black's research on latent heat and thermodynamics played a crucial role in Watt's understanding of the principles governing the behaviour of steam, and Black's insights were instrumental in Watt's later improvements to the steam engine, which was widely adopted in factories during the last quarter of the eighteenth century.

Steam engines powered the industrial revolution, and by

16 Ramsay, Sir William, *The Life and Letters of Joseph Black, M.D.*, Constable and Company Ltd, London, 1918, p 20.

1808 when William Blake published 'And did those feet in ancient time', the 'dark satanic mills' were pumping tonnes of 'fixed air' into the atmosphere. But it was no longer 'fixed air'. In 1787 Antoine Lavoisier took the science of chemistry to a new level with the publication of *Méthode de Nomenclature Chimique* (*Methods of Chemical Nomenclature*)[17] wherein he introduced a systematic way of naming and categorising chemical substances. 'Fixed air' was identified as carbon dioxide, CO_2. But nobody foresaw that the threat of the 'dark satanic mills' was real. Not even Joseph Black.

---

17 Guyton de Morveau, Louis Bernard; Lavoisier, Antoine Laurent; Bertholet, Claude Louis; and Fourcroy, Antoine François, *Méthode de Nomenclature Chimique*, Chez Cuchet, Paris, 1787.

Chapter 3

Jean-Baptiste Joseph Fourier (1768–1830)

It was to be another seventy years before further science related to climate change emerged. Naturally the protagonist had to be a man obsessed with heat. Indeed, Jean-Baptiste Joseph Fourier's life might have been the model for the French gothic novel. Like Joseph Black, he was born in provincial France, and having been orphaned at aged eight, was enrolled in a military school run by the Benedictines of the Convent of St Mark in Auxerre. His first achievements were literary rather than scientific. He wrote sermons delivered to modest acclaim from the pulpits of Paris by prelates who never knew they were written by a twelve-year-old. But later when he was introduced to mathematics, solving polynomial equations became an all-consuming passion. Indeed, in 1789, aged twenty, he was appointed principal chair of mathematics in the Military School of Auxerre.

It was a time of great upheaval, and on July 14th that year Parisians stormed the Bastille. The royal family was forced into virtual imprisonment in the Tuileries Palace. An uneasy peace ensued to the extent that Fourier was able to visit Paris to read his memoir on the resolution of numerical equations of all degrees before the Academy of Sciences. It was well received but stirred controversy over originality. This was not unusual. For example, Darwin's theory of evolution by natural selection has often been referred to as the Darwin–Wallace theory. Given his isolation in Auxerre, it is probable Fourier's work was his own; however, similar ideas had been published earlier by M Budan. Despite this, his visit to Paris had been stimulating and he returned to Auxerre determined to seize the moment.

Before the revolution, French society had been highly stratified. Different social classes had distinct privileges and access to specific occupations. The aristocracy, or nobility, enjoyed significant privileges and held various exclusive positions such as land ownership, military leadership and positions in the church, the judiciary and public administration. Fourier had suffered under these conventions. As a mere citizen before the revolution, he could teach mathematics at the Military School of Auxerre but could not enrol as a student. In the wake of the revolution there would be great change. Revolutionary councils, not just in Paris, encouraged literature, the fine arts and the sciences for all and established an educational framework to make it happen.

But there was a problem. Several European countries sensed a weakness and formed coalitions, launching a series of military campaigns against revolutionary France, thus placing it on a constant war footing. Maths and science teachers who had been teaching in military academies were recruited into artillery units leaving few to nurture France's cultural revolution.

The answer was the École Normale Supérieure (ENS). It was founded in October 1794 by the revolutionary government and initially established in Paris to train teachers for secondary education. The new generation of educators would reinforce revolutionary ideals and help shape the educational system of the newborn republic. Fourier was appointed to the École Normale in 1795. At that time, he was a mathematics professor and a member of the French Academy of Sciences. In the early days he was involved in designing the curriculum, shaping philosophy and selecting faculty. His approach to teaching and his emphasis on mathematical and scientific education made a lasting impact. He believed in the importance of rigorous training in mathematics and the sciences (something we have forgotten in Australia), and his influence contributed to the school's reputation as a centre for advanced scientific and intellectual studies – a reputation that endures to this day.

But his tenure at the École Normale was short lived. In August 1793 a twenty-four-year-old Corsican artillery officer was instrumental in ridding the Port of Toulon of an occupying English fleet. Lieutenant Napoleon Bonaparte was promoted to brigadier general. Two years later, while he was in Paris in hot pursuit of Joséphine de Beauharnais, there was a royalist uprising. The Thermidorian government, which had come to power after the fall of Robespierre, turned to Napoleon for assistance. Napoleon wasn't squeamish about killing French counts. After all, he was middle class Corsican, and it was another opportunity for advancement. As the royalist forces advanced on the Tuileries, Napoleon put an end to their campaign with grapeshot. He left Paris at the head of the French army, marching south to counter a threat from the Austrian Empire. Two years later his victory in the Battle of Rivoli forced the Austrians to accept his terms for peace. The supreme battlefield commander was just twenty-eight years old. Having achieved an uneasy peace in Europe, Napoleon turned his mind to Egypt. Control of Egypt would disrupt English trade with India, but occupation wasn't the only goal. In order to study and document Egypt's history, culture and geography, Napoleon invited a group of scholars and scientists to join the campaign. Joseph Fourier was offered a place. He accepted, and it changed his life.

Napoleon's army landed in Egypt on 1 July 1798, and after the battles of the pyramids and Embabeh, captured Cairo three weeks later. Nelson's defeat of the French navy in the Battle of the Nile the following month was a setback, but despite a communications blackout with France, Napoleon had installed an administration in Cairo by the end of August. By then Fourier had formed more than a nodding acquaintance with the great man who fostered his career in public administration by installing him as French Commissioner on the Divan (Council) of Cairo.

More importantly for posterity, Fourier's stay in Egypt aroused his interest in heat. He became intrigued by the climate of the region and conducted experiments related to the physics of heat. His investigations included the study of heat conductivity through different materials, and his reports made an important contribution to *Description de l'Égypte*,[18] which was published following his return to France in July 1801.

Napoleon had returned to Paris two years earlier to find the latest revolutionary government in disarray. He orchestrated a coup and in November 1799, as First Consul, formed a new administration. He was thirty years old. When Fourier returned from Cairo two years later, Napoleon installed his old comrade as Prefect of the Department of Isère based in Grenoble. It was an inspired appointment, and Fourier, at thirty-three years old, proved a brilliant scientist could be an effective administrator. His welcome in Grenoble was overshadowed by a local spat between republicans and conservatives, represented mainly by the clergy. He declared the Hôtel of the Prefecture neutral ground where both sides could advance their grievances in safety and before long had secured a truce that allowed him to turn to improving infrastructure, agriculture and public services.

Grenoble prospered as Fourier improved its communications and transportation networks. He aligned his prefecture closely with Napoleon's centralising policies, including implementation of the Napoleonic Code and associated legal and educational reforms.

Napoleon made no secret of his support for scientific advancement in France. This was music to Fourier's ears, and in addition to his civic duties he continued his scientific research, particularly in the field of heat diffusion. His studies on the mathematical modelling of heat conduction would later become

18 Commission des Sciences et Arts d'Égypte, *Description de l'Égypte, ou Recueil des observations et des recherches qui ont été faites en Égypte pendant l'expédition de l'armée française*, Imprimerie Impériale, Paris, 1809–1829.

his most famous contribution to science, although much of this work was still in development during his time as prefect.

Fourier's connections within the scientific community and his contributions to the Académie des Sciences kept him active in scientific discourse even while managing his prefectural duties. He also managed to maintain ties with the French Institute in Egypt, which had become highly esteemed in France.

But the idyll couldn't last. In April 1814 Napoleon was forced to abdicate and the Bourbon monarchy under King Louis XVIII was restored. Fourier, like many officials of the Napoleonic regime, had to adjust to the new political environment. He publicly declared his allegiance to the Bourbons and remained in his post, although given his past loyalty to Napoleon his position was precarious.

It became even more precarious when Napoleon escaped from Elba in February 1815 and began his march toward Paris. Officials across France faced a difficult decision: resist Napoleon's return or support him. Grenoble, being a strategic location on Napoleon's route, became a focal point of this dilemma. Fourier, as the prefect of Isère, was in a delicate position. He had to balance his past loyalty to Napoleon with the new Bourbon regime's expectations. Many local soldiers and citizens remained loyal to Napoleon, and the support for his return was growing.

By early March 1815, as Napoleon's advance through southern France gathered momentum, it became clear that resistance in Grenoble would be futile. The city's garrison, along with much of the local population, was sympathetic to Napoleon. For Fourier, it was Hobson's choice; he went with the flow. Napoleon entered Grenoble to great fanfare and without opposition on March 7th. The soldiers and citizens of the city welcomed him enthusiastically, and the troops that had been sent to stop him defected to his side. Fourier, as the senior local official, could do nothing but welcome his old comrade. The exact details of their meeting are not well documented, but one would like to think they had a good lunch and parted as friends.

Napoleon was in Paris two weeks later, and we all know what happened at Waterloo three months after that. It was the end of Fourier's career as a prefect as well, but it did give him time to work on *Théorie Analytique de la Chaleur*,[19] his crowning achievement.

Fourier's primary goal was to understand how heat transfers through solid objects. He worked on developing a mathematical model to describe heat conduction and eventually proposed that any periodic function, such as a repeating pattern of heat distribution, could be represented as an infinite sum of sine and cosine functions. This mathematical method is now known as Fourier transformation and *Théorie Mathématique de la Chaleur* provides a powerful mathematical tool for understanding and describing complex periodic patterns, not only in the context of heat conduction but in many other areas of science and engineering. It was published in June 1822 and laid the foundation for *MEMOIRE sur les Temperatures du Globe Terrestre et des Espaces Planetaires*[20] (MEMOIR on the Temperature of the Earth and Planetary

---

19 Fourier, Jean-Baptiste Joseph, *Théorie Analytique de la Chaleur* (The Analytical Theory of Heat), Firmin-Didot, Paris, 1822.

20 Fourier, Jean-Baptiste Joseph, *Mémoire sur les Températures du Globe Terrestre et des Espaces Planétaires* (Memoir on the Temperatures of the Earth and Planetary Spaces), Académie des Sciences, Paris, 1824.

Spaces), which came five years later. Among other concepts, this memoir advanced the idea that Earth's atmosphere impedes the escape of heat from Earth's surface into space, much like a blanket, thus maintaining a temperature suitable for sustaining life. To quantify the effect of this 'blanket' it would be necessary to determine what Earth's 'equilibrium' temperature would be without it. This temperature is the temperature ($\mathbf{T_e}$) at which the energy Earth receives from the Sun as solar radiation ($\mathbf{E_{in}}$) equals the energy Earth transmits into outer space as infrared radiation ($\mathbf{E_{out}}$).

Fourier's interest in heat was not entirely scientific. On his return from Egypt friends and associates could hardly fail to notice he was 'in the habit of clothing himself, even in the hottest season of the year, after a fashion which is not practised even by travellers condemned to spend the winter amid the snows of the polar regions'.[21] This practice almost certainly precipitated the increasingly frequent fainting attacks he experienced towards the end of his life and the fall down a flight of stairs in May 1830 that killed him.

The first step towards confirming Fourier's 'atmospheric blanket' hypothesis was taken seven years later when French physicist Claude Pouillet developed a pyrheliometer and used it to measure the intensity of solar radiation, known as the solar constant. Pouillet's pyrheliometer was a tube with a thermocouple at its focus. The thermocouple absorbed incoming solar radiation and generated a small electric current proportional to the amount of radiation received. The pyrheliometer was portable and could be pointed directly at

21 Arago, Francois, *Biographies of Distinguished Scientific Men*, Project Gutenberg, September 2005.

the Sun to measure its intensity. Pouillet's measurements led to an estimate for the solar constant of 1228 W/m^2 (watts per square metre). The currently accepted value is 1367 W/m^2.

It is interesting to examine this number in the context of the rooftop solar array on my house. There are sixteen panels with a combined area of 31.8 m^2. Assuming a solar constant of 1367 W/m^2, the theoretical output would be 43.47 kW. In practice, however, it rarely produces more than 5 kW, inferring an efficiency of 11%.

Some of the solar radiation reaching Earth is reflected by cloud and ice so plays no part in warming our planet. This is known as the albedo effect and as a result,

$$\mathbf{E_{in}} = \mathbf{S} \cdot \mathbf{A_p} \cdot (1 - \mathbf{a})$$

where:

- $\mathbf{E_{in}}$ is the total power absorbed per unit surface area, in W/m^2.
- **S** is the solar constant with a value of approximately 1367 W/m^2.
- $\mathbf{A_p}$ is the projected surface area of Earth, in m^2.
- **a** is albedo measured in per cent with an adopted value of 30%.

The reverse is the energy Earth transmits into outer space as infrared radiation. Whereas incoming solar radiation only strikes the half of Earth facing the Sun (the projected surface area of Earth), outbound infrared radiation is transmitted from the entire surface of the planet.

The Stefan–Boltzmann law, which describes the

relationship between the total power radiated by a blackbody and its temperature, was formulated in the late nineteenth century. The law is attributed to the combined work of Josef Stefan and Ludwig Boltzmann and is expressed as:

$$\mathbf{E_{out} = \sigma \cdot A_s \cdot T^4}$$

where:

- $\mathbf{E_{out}}$ is the total power radiated per unit surface area, in W/m^2.
- $\boldsymbol{\sigma}$ is the Stefan–Boltzmann constant, which is a fundamental physical constant with a value of approximately 5.67×10^{-8} $W/(m^2 \cdot K^4)$.
- $\mathbf{A_s}$ is the surface area of the object, in m^2.
- $\mathbf{T}$ is the absolute temperature of the object, in degrees kelvin (K).

Since $\quad \mathbf{E_{in} = E_{out}}$

$$\mathbf{T^4 = S \cdot A_p \cdot (1\text{-}a)/\sigma \cdot A_s}$$

$\mathbf{A_p/A_s}$ = ¼, therefore

$$\mathbf{T = ((1\text{-}a)S/4\sigma)^{0.25}}$$

= 255 K or -18°C

The pre-industrial global average temperature referenced in the Paris Agreement (see page 35) refers to the period before widespread industrialisation began in the mid-eighteenth century, specifically around the year 1750, and is considered to have been between 13.7 and 14.0 degrees Celsius. Without greenhouse gases in the atmosphere, Earth would lose heat more rapidly, and its average surface temperature would not be far from the - 18°C estimated above.

So, as Joseph Fourier suspected, Earth's atmosphere plays a crucial role in warming the planet. We now know the Sun emits a broad spectrum of electromagnetic radiation, including visible light. This solar radiation passes through Earth's atmosphere and some of it is absorbed, warming the surface, while some of it is reflected into space by cloud and ice. This is the albedo effect mentioned above. As Earth's surface warms, it emits thermal radiation in the form of infrared radiation. This outgoing infrared radiation carries energy away from the surface into the atmosphere. Certain gases in Earth's atmosphere, now known as greenhouse gases, have the property of absorbing and re-emitting infrared radiation. These gases include water vapour (H_2O), carbon dioxide (CO_2), methane (CH_4), nitrous oxide (N_2O) and others. When they absorb energy, they become energised and re-emit infrared radiation in all directions, including back towards Earth's surface.

This re-emission of infrared radiation by greenhouse gases effectively traps heat in the lower atmosphere, preventing it from escaping directly into space. This trapped heat warms the lower atmosphere and, in turn, warms Earth's surface. As Fourier foretold, it acts like a blanket, keeping the planet's surface warmer than it would be if all the heat escaped directly into space.

The importance of greenhouse gases in terms of their overall heating power depends on several factors, including their concentration in the atmosphere, their radiative properties and their lifetimes. Water vapour is by far the most abundant greenhouse gas in Earth's atmosphere and has the greatest overall heating power. However, its concentration is primarily controlled by temperature and varies significantly with location and time. Of the runners up, nitrous oxide and methane are much more potent than CO_2 in terms of heating power; however, since there is far more CO_2 in the atmosphere than either methane or nitrous oxide the heating powers of the trio are of similar magnitudes.

But none of this was understood until 1856 when a housewife from Seneca Falls, New York, was the first to make the connection between CO_2 and climate change. She suggested increased atmospheric concentration would raise temperatures on Earth. Nobody took much notice. Perhaps she was the wrong sex.

Chapter 4

Eunice Newton Foote (1819–1898)

Full many a gem of purest ray serene
The dark unfathom'd caves of ocean bear:
Full many a flower is born to blush unseen,
And waste its sweetness on the desert air.[22]

Eunice Newton Foote may have been born to blush unseen, but despite her modest origins and her rudimentary grounding in science she was undoubtedly the first to suggest:

> An atmosphere of that gas [carbon dioxide] would give to our earth a high temperature; and if as some suppose, at one period of its history the air had mixed with it a larger proportion than at present, an increased temperature from its own action as well as from increased weight must have necessarily resulted.[23]

22 Gray, Thomas, 'Elegy Written in a Country Churchyard', 1751.
23 Foote, Eunice, Circumstances Affecting the Heat of the Sun's Rays, *The American Journal of Science and Arts*, 1856, vol. 22, no. 66, pp 382–383.

Eunice was born in Goshen, Connecticut, in the summer of 1819. It's been suggested that her father, Isaac, was related to the legendary Sir Isaac, but even if he were he couldn't have been a direct descendent since the great man never married and, as far as we know, had no children. Nevertheless, her father undoubtedly had a reverence for science since although Eunice was one of seven sisters and five brothers, she was sent to the Troy Female Seminary, the only school for girls in the United States that offered a scientific curriculum and a chemistry laboratory.

She graduated from the seminary in 1838, and although we know little about her salad days, she could not have gone to a university since that was not for women until much later. Yale University admitted its first female students to its School of Fine Arts in 1870, and Stanford University didn't become coeducational until 1892. But we do know she married Elisha Foote Jr in 1841, and since Elisha had been appointed district attorney for Seneca County, they went to live in Seneca Falls in the Finger Lakes region of upstate New York.

Seneca Falls may have been remote, but being surrounded by rich agricultural land with abundant precipitation and on the banks of a fast-flowing river, it attracted farmers and industrialists alike. There were no dark satanic mills. The river provided more than enough power to drive the flour, saw and wool mills crowding its banks. As prosperous as it was, however, the town of four thousand or so was not without its divisions. The 1776 Declaration of Independence affirmed 'all men are created equal and are endowed by their Creator with certain unalienable rights, including life, liberty, and the pursuit of happiness'. Abolitionists argued this applied to all men, including slaves, and Seneca Falls with its formidable Quaker presence had become an important hub for the abolitionist movement. Fugitive slaves on their way to Canada were often sheltered in the town's basements and attics.

Daughters Mary and Augusta followed in rapid succession, but those momentous events didn't dampen Eunice's passion for science or her new interest, invention. Elisha, despite being a lawyer, was an amateur inventor and had persuaded Eunice her grounding in science would give her a head start in that pursuit. She took up the challenge and, having moved into their new house, designed, installed and equipped a laboratory. Basic apparatus and equipment at the time would have included simple single-lens microscopes; balances and scales; mercury-in-glass thermometers; optical equipment such as prisms, lenses and mirrors; beakers, flasks, test tubes and graduated cylinders; Bunsen burners fuelled by coal gas; a vacuum pump and electrical equipment, including a Leyden jar, galvanic cells and a gold leaf electrometer.

The paint on Eunice's new laboratory could barely have dried before she patented her first commercial invention, a thermostatically controlled cooking stove. In this case, necessity was truly the mother of invention. Only a young wife struggling to establish a new household and expecting a new baby could have appreciated just how much time she spent monitoring and regulating the temperature of her cooking stove. Her idea, simply outlined in a single sentence in the letter patent dated 26 May 1842, was as follows:

> My plan makes the stove or other structure in which it may be used regulate its own heat and this is effected by applying the expansive and contracting power of a metallic rod by different degrees of heat to the damper of the stove by which the admission of air thereto is governed so that when the heat shall rise above any required degree the expansion shall close such damper and when it shall fall below such degree the contraction shall open it and thus keep a uniform heat of the requisite intensity.[24]

The patent was registered in Elisha's name since it would not have been enforceable under her own. It must have been a commercial success since in 1849 in the case of *Foote v Silsby*, Elisha won $1500 (about $200 000 now) for patent infringement.

Perhaps that's why she involved herself in the women's rights movement, a movement fostered by abolitionist thinking. The members argued that the founding principle of the Declaration of Independence applied to women as well as slaves and were agitating for suffrage, rights to property ownership and access to tertiary education. These were issues dear to Eunice Newton Foote's heart and not too far from her doorstep either since her nearest neighbour and fellow newcomer to Seneca Falls was Elizabeth Cady Stanton, the movement's leading thinker and campaigner. In 1848 Stanton organised the first women's rights convention, not in New York, not in Boston, not in Washington, but in Seneca Falls. The number one item on the convention agenda was ratification of

24 Foote, Elisha Jr, *Regulating the Draft of Stoves*, Specification of Letters Patent No. 2636, United States Patent Office, 1842.

the 'Declaration of Rights and Sentiments' she had published expressly for that purpose. Three hundred attended, of which sixty-eight women and thirty-two men, including Eunice and Elisha Foote, signed the declaration.

They must have made some progress because by 1860 when Eunice patented her next invention, an insert of vulcanised rubber to 'prevent the squeaking of boots and shoes', she did so in her own name.

But despite her successful inventions Eunice's mind was never too far away from science. The American Association for the Advancement of Science (AAAS) held its first meeting in Philadelphia in the fall of 1948, and its subsequent publications sought to keep members informed of science's latest developments. The hot topic at the time within the geological community was transitioning from the Plutonist to the uniformitarianism theory of geological development. Plutonists believed igneous processes shaped Earth's surface. Uniformitarians believed the natural processes we observe today, such as erosion, sedimentation and volcanic activity, have been shaping Earth's surface over geological time, and these processes have operated uniformly throughout Earth's history. Charles Lyell's book *Principles of Geology*, first published in 1830, popularised this idea. It also suggested the lush vegetation and extensive coal deposits of the Carboniferous period were a result of a warm and humid climate influenced by elevated levels of atmospheric CO_2.

Another issue of the day was why, in alpine regions, it was warmer in the valleys than on the mountain tops. Some argued it was the density of the atmosphere. Others argued lower temperatures at altitude were caused by reduced incident angles of the Sun's rays on mountain slopes.

Eunice provided answers to these questions by placing glass cylinders filled with different gases at different pressures in the sunlight and recording and comparing the temperature rises in each cylinder. After ten minutes the temperature in a cylinder filled with common air rose 5.6ºC while that in a cylinder filled with CO_2 rose 16.7ºC. She also demonstrated that it was colder at the top of a mountain than at the bottom of a valley below by placing cylinders filled with low pressure air and compressed air side by side in the sun for twelve minutes. The temperature in the low-pressure air cylinder rose 4.4ºC while that in the compressed air cylinder rose 16.7ºC. These were interesting findings; however, Eunice went a step further and that is why she is now referred to as the mother of climate change. Her masterstroke was predicting that a future increase in atmospheric CO_2 levels would warm the planet.

Her paper, 'Circumstances Affecting the Heat of the Sun's Rays', was presented at the tenth annual meeting of the AAAS held in Albany, New York, in August 1856. Eunice may have been able to register a patent in her own name four years later, but in 1856 the gentlemen of the AAAS were not quite ready to extend a welcome to an amateur scientist who happened to be a woman as well. Her paper was presented on her behalf by Joseph Henry, secretary of the Smithsonian Institution, who opined, 'Science was of no country and of no sex. The sphere of woman embraces not only the beautiful and the useful, but the true'.[25] Saucy devil.

25 Mandel, Kyla, 'This Woman Fundamentally Changed Climate Science—And You've Probably Never Heard of Her', ThinkProgress, Washington DC, Center for American Progress Action Fund, 18 May 2018.

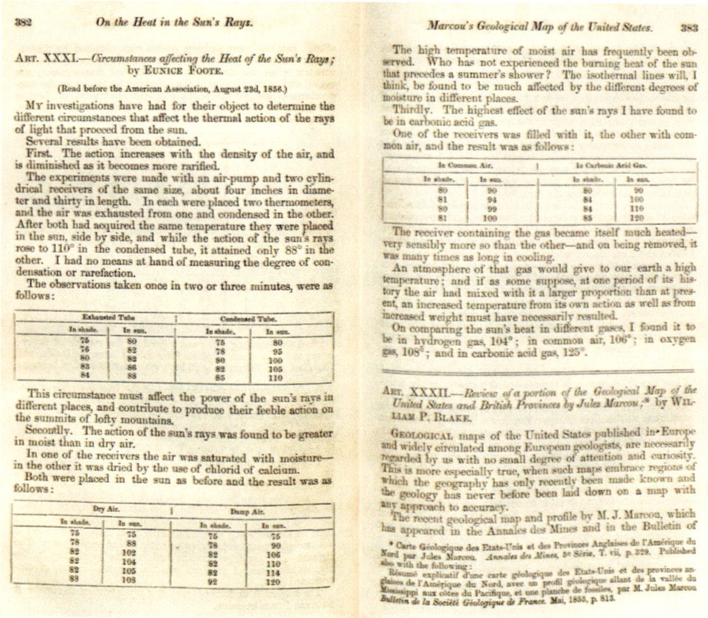

382 *On the Heat in the Sun's Rays.*

ART. XXXI.—*Circumstances affecting the Heat of the Sun's Rays;* by EUNICE FOOTE.

(Read before the American Association, August 23d, 1856.)

MY investigations have had for their object to determine the different circumstances that affect the thermal action of the rays of light that proceed from the sun.

Several results have been obtained.

First. The action increases with the density of the air, and is diminished as it becomes more rarified.

The experiments were made with an air-pump and two cylindrical receivers of the same size, about four inches in diameter and thirty in length. In each were placed two thermometers, and the air was exhausted from one and condensed in the other. After both had acquired the same temperature they were placed in the sun, side by side, and while the action of the sun's rays rose to 110° in the condensed tube, it attained only 88° in the other. I had no means at hand of measuring the degree of condensation or rarefaction.

The observations taken once in two or three minutes, were as follows:

Exhausted Tube		Condensed Tube.	
In shade.	In sun.	In shade.	In sun.
75	80	75	80
76	82	78	95
80	82	80	100
83	86	82	105
84	88	85	110

This circumstance must affect the power of the sun's rays in different places, and contribute to produce their feeble action on the summits of lofty mountains.

Secondly. The action of the sun's rays was found to be greater in moist than in dry air.

In one of the receivers the air was saturated with moisture—in the other it was dried by the use of chlorid of calcium.

Both were placed in the sun as before and the result was as follows:

Dry Air.		Damp Air.	
In shade.	In sun.	In shade.	In sun.
75	75	75	75
78	88	78	90
82	102	82	106
82	104	82	110
82	105	82	114
83	108	92	120

Marcou's Geological Map of the United States. 383

The high temperature of moist air has frequently been observed. Who has not experienced the burning heat of the sun that precedes a summer's shower? The isothermal lines will, I think, be found to be much affected by the different degrees of moisture in different places.

Thirdly. The highest effect of the sun's rays I have found to be in carbonic acid gas.

One of the receivers was filled with it, the other with common air, and the result was as follows:

In Common Air.		In Carbonic Acid Gas.	
In shade.	In sun.	In shade.	In sun.
80	90	80	90
81	94	84	100
80	99	84	110
81	100	85	120

The receiver containing the gas became itself much heated—very sensibly more so than the other—and on being removed, it was many times as long in cooling.

An atmosphere of that gas would give to our earth a high temperature; and if as some suppose, at one period of its history the air had mixed with it a larger proportion than at present, an increased temperature from its own action as well as from increased weight must have necessarily resulted.

On comparing the sun's heat in different gases, I found it to be in hydrogen gas, 104°; in common air, 106°; in oxygen gas, 108°; and in carbonic acid gas, 125°.

ART. XXXII.—*Review of a portion of the Geological Map of the United States and British Provinces by Jules Marcou;** by WILLIAM P. BLAKE.

GEOLOGICAL maps of the United States published in Europe and widely circulated among European geologists, are necessarily regarded by us with no small degree of attention and curiosity. This is more especially true, when such maps embrace regions of which the geography has only recently been made known and the geology has never before been laid down on a map with any approach to accuracy.

The recent geological map and profile by M. J. Marcou, which has appeared in the Annales des Mines and in the Bulletin of

* Carte Géologique des Etats-Unis et des Provinces Anglaises de l'Amérique du Nord par Jules Marcou. *Annales des Mines*, 5e Série, T. vii, p. 329. Published also with the following:

Résumé explicatif d'une carte géologique des Etats-Unis et des provinces anglaises de l'Amérique du Nord, avec un profil géologique allant de la vallée du Mississippi aux côtes du Pacifique, et une planche de fossiles, par M. Jules Marcou *Bulletin de la Société Géologique de France.* Mai, 1855, p. 813.

Figure 4.1 'Circumstances Affecting the Heat of the Sun's Rays' (Photo: Wikimedia Commons[26])

The American Journal of Science and Arts published the paper in Eunice's name three months later.[27] It was a first for American women. Her friend and neighbour Elizabeth Cady Stanton would have been delighted. But women's rights and climate science were soon to be off the agenda. Following Lincoln's victory in the 1860 presidential race, Southern states, alarmed by his position on slavery and fearing his administration would threaten their economy and way of life,

26 https://commons.wikimedia.org/wiki/File:Eunice_Foote_%22Circumstances_Affecting_the_Heat_of_the_Sun%27s_Rays%22_(1856).png, accessed November 2024.

27 Foote, Eunice, Circumstances Affecting the Heat of the Sun's Rays, *The American Journal of Science and Arts*, 1856, vol. 22, no. 66, pp 382–383.

began to secede from the Union. South Carolina was the first on 20 December 1860.

Others soon followed and as tensions between North and South escalated the Confederates demanded the surrender of Fort Sumter, still under Union control in Charleston Harbor. When the Union refused, the Confederate Army under General PGT Beauregard began bombarding it. It was 12 April 1861, and it was the beginning of the American Civil War. Americans had more to worry about than global warming, and Eunice's groundbreaking contribution might have been forgotten but for another informed amateur, a retired petroleum geologist with a fascination for old scientific publications.

In 2011, while trawling through back issues of the *American Journal of Science and Arts*, Raymond P Sorenson stumbled on Eunice's seminal paper from 1856. Being a geologist with an interest in climate science, he knew enough to know that Irishman, John Tyndall, who had been recognised as the first to link atmospheric CO_2 levels to climate change, did not publish his work 'On the Absorption and Radiation of Heat by Gaseous and Liquid Bodies' until 1861. His papers explained the role of atmospheric gases in regulating Earth's temperature and established the basic science of the greenhouse effect, now widely accepted as the mechanism behind global warming and climate change. Just as Eunice Newton Foote had done five years earlier.

Nothing stirs the cosy cloisters of academia more than a brawl over priority. But it doesn't always have to end in tears. Jean Baptiste Joseph Fourier readily accepted that Ferdinand François Désiré Budan de Boislaurent's 1807 theorem for the analysis of polynomial equations was ahead of his time. However, academia in general appears to accept Fourier's and Boislaurent's theorems were developed independently. Another example is that Charles Darwin had been working on his theory of evolution through natural selection for over two decades but had yet to publish his ideas when, in June 1858, he received a letter from Alfred Wallace outlining a similar concept. Darwin resolved the conflict by presenting Wallace's essay along with his own unpublished writings at a meeting of the Linnean Society of London on 1 July 1858.

There can be no doubt Eunice Newton Foote was the first to make the connection between CO_2 and climate change. Tyndall's thesis may have been more rigorous, but it was five years late and didn't acknowledge Eunice Newton Foote. Perhaps he wasn't aware of her paper. It's a hot debate, and there are plausible arguments on both sides but, if he was aware of it, why did he not recognise her? Because she was an amateur, because she was a colonial or because she was a woman? We will never know, but the next hero in the annals of climate science owed it all to a woman. Without his wife's courage it's unlikely his work would ever have started.

Chapter 5

Milutin Milanković (1879–1958)

Milutin Milanković was born in the village of Dalj on the west bank of the Danube in present-day Croatia on 28 May 1879. His family was comfortable, well-educated and Serbian, but he suffered from poor health so received his primary education at home. His health improved, so for his secondary education he was able to attend the Royal Grammar School in Osijek, where he excelled in maths and science. He matriculated in 1896 with honours and was accepted at Vienna University of Technology, one of the leading technical universities in Europe, where he studied civil engineering with a focus on structural engineering and construction. He graduated in 1902 but must have enjoyed university life since he spent a further two years earning a doctorate in concrete research. On completing the doctorate, Milanković remained in Vienna working as a civil engineer, successfully applying his novel techniques to the design and construction of alpine dam and bridge structures to the extent that by 1909, when he accepted a lectureship in applied mathematics at the University of Belgrade, he was a wealthy man.

He began teaching courses in applied mathematics, mechanics and theoretical physics, but the role provided him with the freedom to pursue his research interests, and once established at the university Milanković gradually shifted his focus from engineering to his research on the impact of astronomical factors on Earth's climate. He started developing his theories on climate cycles and the effects of changes in Earth's orbit and axial tilt on long-term climate patterns.

But that wasn't Milutin's only universe. Serbia had gained full independence from the Ottoman Empire at the Congress of Berlin in 1878, so the early 1900s in Belgrade were a period of significant cultural development with a strong movement to revive and promote Serbian culture, language and traditions. Intellectuals, writers and artists played a key role in this renaissance and Milutin Milanković was in the thick of it. Perhaps that's how he met Kristina Topuzović. 'Tinka' came from a well-to-do Serbian family, so they were the perfect match. They married in Belgrade on 24 June 1914 and four days later were on honeymoon in Milutin's hometown, Dalj. That's when three Bosnian Serb nationalists murdered the Austrian Archduke Franz Ferdinand and his wife in Sarajevo. Four days later Austria–Hungary declared war on Serbia and so began World War One. It was also the beginning of Milanković's thirty-year crusade to explain the rhythm of the ice ages.

The crusade had a bumpy start. In June 1914 Dalj was part of the Kingdom of Croatia–Slavonia, an autonomous kingdom within the Austro–Hungarian Empire, so even though it was

Milanković's birthplace he was arrested as an enemy alien and sent to an internment camp in Nezsider, Hungary (today Neusiedl am See, Austria). That might well have been the end of our story but for the poise and bravery of Milanković's bride who somehow made her way to Vienna to ask Milutin's former teacher, Professor Emanuel Czuber, to intercede on her husband's behalf. Czuber was an advisor to the Austrian court and Archduke Ferdinand Karl of Austria's father-in-law, so was well connected; however, it was his son-in-law's brother who had been assassinated in Sarajevo, so he must have been placed in a difficult position. Nevertheless, he secured a parole for his former student who was allowed to settle in Budapest with his bride as a guest of the Hungarian Academy of Sciences for the next four years.

Milanković didn't publish his *Canon of Insolation and the Ice-Age Problem* until 1941; however, the project was undoubtedly well under way by the end of World War One. He identified and modelled cycles of long-term changes in Earth's climate caused by variations in the way our planet orbits the Sun. These cycles explain why Earth experiences ice ages and warmer periods over tens of thousands to hundreds of thousands of years. There are three main Milanković cycles.

1. The Eccentric is a 100 000-year cycle tracing Earth's orbit around the Sun from being almost circular to more oval-shaped and back again. When the orbit is more oval, Earth is sometimes farther from the Sun, leading to cooler periods. When the orbit is more circular, the distance from the Sun is more constant, leading to milder climate variations.

2. The Obliquity cycle is a 41 000-year cycle caused by changes in the angle of Earth's tilt on its axis between 22.1 degrees and 24.5 degrees. A greater tilt means more extreme seasons – hotter summers and colder winters. A smaller tilt means milder seasons – cooler summers and warmer winters. These changes in tilt affect how much sunlight different parts of Earth receive.

3. The last cycle is described as Precession. Earth's axis wobbles like a spinning top, completing a full wobble approximately every 26 000 years. This wobble changes the timing of the seasons in relation to Earth's position in its orbit. Over time, this can significantly alter the climate, making winters in one hemisphere warmer and summers cooler, and vice versa.

These cycles interact with each other and can align in ways that either amplify or reduce their effects on Earth's climate. Certain alignments lead to the onset of ice ages and others result in interglacial or warmer periods. By studying these cycles, scientists can understand past climate patterns and predict future changes.

We are currently in an interglacial period. Earth's orbit is close to its least elliptical (most circular) state, meaning the difference in distance from the Sun over the course of a year is minimised leading to relatively stable climate conditions in terms of solar radiation. Earth's axial tilt is about 23.4 degrees and is moving towards the lower end of the range (22.1 degrees) in the 41 000-year cycle, leading to less extreme seasonal variations.

Regarding 'Precession', Earth's axis is currently such that the Northern Hemisphere experiences winter when Earth is closest to the Sun (perihelion) and summer when it is farthest from the Sun (aphelion). This configuration reduces the contrast between summer and winter in the Northern Hemisphere, while increasing it in the Southern Hemisphere. Over the next 13 000 years this will reverse, with Northern Hemisphere summers occurring at perihelion.

The combined effect is that we are now in an interglacial period with relatively mild and stable climate conditions. It has been named the Holocene epoch and it began about 11 700 years ago. The natural cycles indicate that another ice age could be expected in the future; however, human-induced climate change is significantly impacting these natural patterns, leading to who knows where.

In 1919, with World War One over, Milutin and Tinka with their four-year-old son, Vasko, returned to Belgrade to resume his work at the university. In 1941, following the Nazi German invasion of Yugoslavia and occupation of its capital, Milanković retreated to his home and continued his research. He resumed his work as a university professor when the war ended, retired in 1955 and died three years later, aged 79.

It was to be another twenty years before scientists had the tools, the techniques and the money needed to drill deep into the ocean floor and ice sheets, allowing them to extract long cores containing layers of sediment and ice that recorded Earth's past climate. They could accurately date these layers and analyse the chemical composition to understand temperature and atmospheric changes over time.

These advances enabled Hays, Imbrie and Shackleton to show that Earth's periodic ice ages were closely linked to variations in Earth's orbit around the Sun, thus confirming Milanković's calculations with empirical evidence.

The work was based on an analysis of two continuous cores, 9.54 and 14.59 metres in length, each taken in the Southern Ocean in waters over three kilometres deep and representing periods of three and five million years, respectively. Their discoveries were published by the AAAS in 1976 in the aptly titled paper, 'Variations in the Earth's Orbit: Pacemaker of the Ice Ages'.[28]

Apart from confirming Milanković's theory that orbital variations are the primary drivers of ice ages, the paper confirmed the role of CO_2 in a critical feedback mechanism for climate processes. Subsequent research showed that CO_2 levels amplify the climate changes initiated by orbital variations. Thus, while orbital variations trigger ice ages, CO_2 and other factors significantly influence their magnitude and duration. One of the pioneers of this thinking was Charles David Keeling whose groundbreaking work has inspired generations of climate scientists and remains a cornerstone of climate research and policy.

---

28 Hays, James D; Imbrie, John; and Shackleton, Nicholas J, Variations in the Earth's Orbit: Pacemaker of the Ice Ages, *Science*, vol. 194, no. 4270, 1976, pp 1121–1132. https://doi.org/10.1126/science.194.4270.1121.

Chapter 6

Keeling and Co

Charles David Keeling earned his PhD in chemistry from Northwestern University in 1954. During his postdoctoral research at the California Institute of Technology he developed an interest in atmospheric CO_2 and in 1956 received a postdoctoral fellowship from the National Research Council to continue his studies on CO_2 concentrations. He began developing methods for precise CO_2 measurements using manometric techniques but soon recognised the potential of infrared gas analysers, which could be adapted for continuous and precise long-term measurements.

His work attracted the attention of Roger Revelle, then the director of Scripps Institution of Oceanography and a prominent figure in climate research. Recognising the importance of Keeling's research, Revelle invited him to join Scripps in 1956 and provided the necessary support and resources for Keeling to establish a systematic program for measuring atmospheric CO_2 levels. Keeling accepted; however, early measurements in Big Sur, California, experienced significant daily variations reinforcing the need for continuous long-term monitoring in a remote location.

Fortunately, post-World War Two, the US Weather Bureau took the initiative to enhance weather forecasting and atmospheric science by establishing high-altitude observatories for atmospheric monitoring in remote locations. Mauna Loa was the first. It was in the middle of the Pacific, so definitely remote, and it was above the cloud line on top of a 3500-metre-high volcano. So, as long as the volcano was dormant, the conditions were ideal for obtaining uncontaminated atmospheric data. The initial infrastructure was ready in 1956. Keeling began his groundbreaking CO_2 measurements in 1958 and so started the now-famous Keeling Curve.

Keeling's findings supported the understanding that CO_2 acts as part of a critical feedback mechanism in climate processes. As CO_2 levels rise, the greenhouse effect intensifies, leading to higher global temperatures. Higher temperatures, in turn, can lead to increased CO_2 release from natural sources (e.g. permafrost thaw, forest fires), creating a feedback loop that amplifies climate change. The Keeling Curve vividly illustrates these concepts (see Figure 6.1). The seasonal variations are driven by the natural processes of photosynthesis and respiration in plants, primarily in the Northern Hemisphere. During the growing season, plants absorb CO_2 thus decreasing atmospheric CO_2 levels, while in the autumn and winter, the reduced photosynthetic activity and increased respiration and decay release CO_2 back into the atmosphere, causing an increase in CO_2 levels. This cyclical pattern is superimposed on the long-term trend of rising atmospheric CO_2 due to human activities. Given that 1 ppm of atmospheric CO_2 weighs 7.821 Gt, a seasonal swing of 6 ppm equates to a movement of 47 Gt.

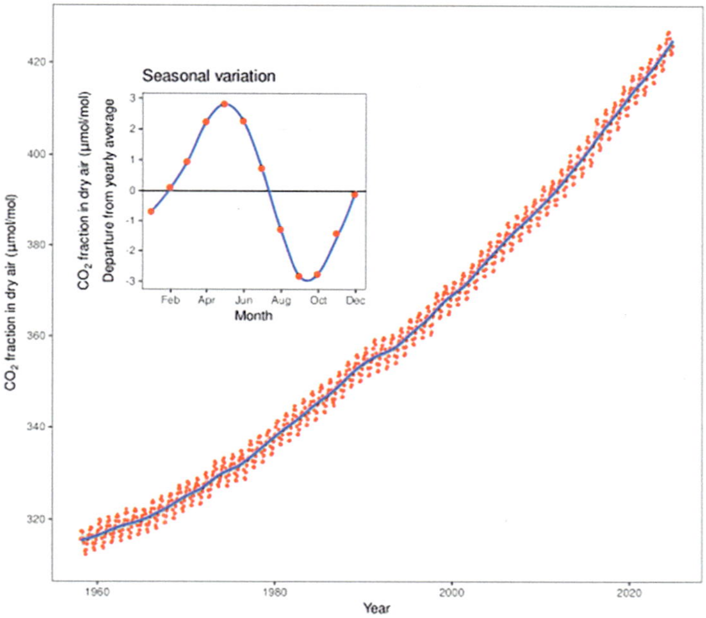

Figure 6.1 The Keeling Curve. Atmospheric carbon dioxide (CO_2) concentrations, as monthly mean, from 1958 to 2023[29]

While Charles Keeling was measuring and recording the rise in atmospheric CO_2 concentration, his colleague Roger Revelle was studying the capacity of the oceans to absorb it. He found that while the oceans can absorb a substantial amount of CO_2, their capacity is limited and certainly not up to accommodating the rapid increase in atmospheric CO_2 levels resulting from

29 Tans, Dr Pieter (NOAA/ESRL); and Keeling, Dr Ralph (Scripps Institution of Oceanography), Monthly mean CO_2 concentration, Mauna Loa 1958–2023, https://commons.wikimedia.org/wiki/Template:Other_versions/Mauna_Loa_CO2_monthly_mean_concentration, accessed 15 December 2023.

human activities without significant 'acidification' (reduction in pH). A review of the historical data (Table 7.1, see later discussion, chapter 7) indicates that, in the wake of the industrial revolution, the oceans began absorbing CO_2 in 1912. Since then, the accumulated absorption has been 397 Gt of CO_2. This may be a drop in the ocean, but the effects are palpable.

Pre-industrial ocean pH is estimated to have been relatively stable at about 8.2. This value represents a slightly alkaline environment, which is typical for sea water. The start of the industrial revolution marked the beginning of increased CO_2 emissions, but significant changes in ocean pH were not yet evident until between 1900 and 1950 when average pH dropped to around 8.15. The mid-twentieth century saw a significant increase in industrial activities and fossil fuel consumption, resulting in a more noticeable decrease in ocean pH to 8.1. The early twenty-first century has seen further declines to the present-day level of 8.05, reflecting continued high levels of CO_2 emissions.

The process by which rising atmospheric CO_2 concentrations lead to ocean acidification is primarily driven by a series of chemical reactions between CO_2 and sea water. As atmospheric CO_2 concentrations increase, more CO_2 is absorbed by the oceans. CO_2 dissolves in sea water and forms carbonic acid (H_2CO_3) through the following reaction:

$$CO_2 + H_2O \rightarrow H_2CO_3$$

This carbonic acid is relatively weak and dissociates (breaks down) into bicarbonate (HCO_3^-) and hydrogen ions (H^+):

$$H_2CO_3 \rightarrow HCO_3^- + H^+$$

The increase in hydrogen ions is the key driver of ocean acidification. The concentration of H^+ in water determines its acidity, so as more hydrogen ions are produced, the pH of sea water decreases, making it more acidic.

The increase in hydrogen ions also affects the concentration of carbonate ions (CO_3^{2-}) in the ocean. The excess hydrogen ions react with carbonate ions to form bicarbonate, reducing the availability of carbonate ions:

$$H^+ + CO_3^{2-} \rightarrow HCO_3^-$$

The decline in pH from 8.2 to 8.05 represents a 30% increase in the hydrogen ion concentration, indicating significant acidification. This is unprecedented in the geological record, raising concerns about the impact on marine ecosystems and organisms, particularly those that rely on calcium carbonate for their shells and skeletons. The problem is, even if net zero is achieved by 2050, the accumulated oceanic CO_2 absorption will have grown from 397 to 693 Gt (Table 7.1) – an inconvenient case of acid tummy.

Roger Revelle left Scripps in 1963 to join the Harvard University faculty as director of the Center for Population Studies. His work there included research on the interactions between population, resources and the environment and was built on his extensive background in oceanography and climate science. One of his students, Al Gore, was the son of a senator from Tennessee. Gore's main academic pursuit was in the field of government where he focused on political science, policy analysis and the impact of media on politics. He also attended Roger Revelle's course on environmental science, and this played a crucial role in shaping Gore's future career in politics and environmental advocacy.

After graduating from Harvard with a Bachelor of Arts degree in government in 1969, Gore enlisted and in 1971 spent five months in Vietnam as a military journalist. Afterwards he worked as a reporter for *The Tennessean*, a newspaper in Nashville, Tennessee, before taking leave of absence to attend Vanderbilt University Law School. He didn't complete the course. In 1976 his father's former seat in the House of Representatives fell vacant, so he ran for it and won.

Gore was in the House of Representatives until 1985, the Senate until 1993 and was vice president to Bill Clinton from 1993 to 2001. During those years he was an energetic and effective advocate for environmental issues. In 1976 he organised the first congressional hearings on climate change, focusing on the potential impact of greenhouse gases on global temperatures. He worked with the Tennessee Valley Authority to ensure it adhered to environmental regulations and pushed for the modernisation of its facilities to reduce pollution.

As a senator he sponsored the *Global Climate Protection Act 1987*, which directed the US Government to formulate a national policy on global warming and to increase research on the effects of greenhouse gases. In 1990 he played a significant role in promoting further amendments to the *Clean Air Act Amendments 1970* aiming to reduce air pollution and acid rain.

Gore was a key participant in the UN-sponsored Earth

Summit in Rio de Janeiro in 1992. As discussed in chapter 7, this was the summit that culminated in the treaty with no mandatory emission reduction targets. Five years later, Gore, as US vice president, was instrumental in achieving the targets set at Kyoto. But before doing so he published *Earth in the Balance: Ecology and the Human Spirit,*[30] foreshadowing the global environmental crisis and proposing solutions to address it.

Earth in the Balance presented a holistic view of the environmental crisis, attributing it to a combination of industrialisation, unsustainable practices and a lack of awareness. His proposed solutions encompassed international cooperation, policy reform, economic incentives, technological innovation and a fundamental shift in human values and behaviour. He called for urgent and comprehensive action to address the environmental challenges and ensure a sustainable future for the planet, and indeed, *Earth in the Balance* was the blueprint for Gore's role as VP.

The man destined to be the next president of the United States lost the November 2000 election to George W Bush in one of the most contentious and closely contested elections in US history. The pivotal moment was the Supreme Court's decision in *Bush v Gore*, which halted the Florida recounts and awarded the state's electoral votes to Bush, thus securing his victory. Despite winning the popular vote, Gore's loss in the Electoral College highlighted the complexities and potential issues within the US electoral system.

Perhaps everything really is for the best in the best of all possible worlds. The loss must have been a bitter moment for

30 Gore, Al, *Earth in the Balance: Ecology and the Human Spirit*, Houghton Mifflin, Boston, 1992.

Gore, but he picked himself up and returned to environmental activism with the release of the highly acclaimed documentary film *An Inconvenient Truth*. It was a roaring success and won an Academy Award for best documentary feature. Gore won the Nobel Peace Prize in 2007, which he shared with the IPCC.

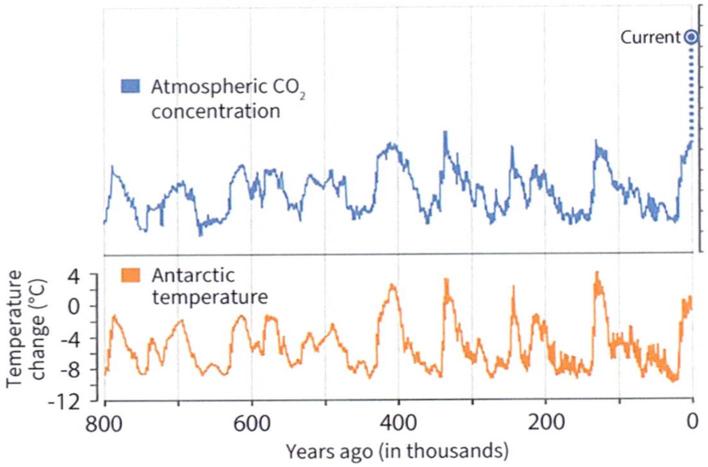

Figure 6.2 Antarctic temperatures and atmospheric CO_2 concentrations over the past 800 000 years[31]

The key moment in Gore's film is when he compares the CO_2 and ice cycles, which for hundreds of thousands of years have been in sync with regular amplitudes of circa 100 ppm CO_2 and 10°C. It's been established that, hitherto, the ice cycle was determined by Milanković cycles and if there were no other changes we would be heading into a cold cycle. But this is no longer a given. Within less than one hundred years

31 Antarctic temperatures and atmospheric CO_2 concentrations over the past 800 000 years. Figure by Jeremy Shakun, The Royal Society; data from Lüthi et al, *Nature*, 2008, vol. 453, pp 379–382, and Jouzel et al, *Science*, 2007, vol. 317, pp 793–796.

the CO_2 cycle amplitude has doubled to more than 200 ppm and the global average land–sea temperature anomaly (mean temperature difference from the 1961–1990 average, in °C) has risen to 1.10°C. As the amateur scientist from upstate New York had suggested 150 years earlier,

> An atmosphere of that gas [carbon dioxide] would give to our earth a high temperature; and if as some suppose, at one period of its history the air had mixed with it a larger proportion than at present, an increased temperature from its own action as well as from increased weight must have necessarily resulted.[32]

32 Foote, Eunice, Circumstances Affecting the Heat of the Sun's Rays, *The American Journal of Science and Arts*, 1856, vol. 22, no. 66, pp 382–383.

Chapter 7

Net Zero 2050

Since 1912 the world's population has grown from 1.8 to 8.2 billion. In the process it has burnt 600 to 800 billion tonnes (Gt) of coal, oil and natural gas, thereby releasing 1.8 trillion tonnes of CO_2 into the atmosphere. One trillion tonnes of it have remained, trapping solar energy and raising Earth's temperature by 1.10°C (ΔT). This phenomenon is referred to as global warming. The rest, around 800 Gt, has been absorbed by land and sea in roughly equal proportions. The CO_2 taken in by the countryside has been largely beneficial, protecting biodiversity and enhancing soil fertility. That taken in by the oceans, as outlined in the previous chapter, has not been so desirable. It has reduced background pH in the oceans from 8.2 to 8.05, a problem referred to as ocean acidification.

Following publication of the IPCC's First Assessment Report in 1990, the UN convened the United Nations Conference on Environment and Development in Rio de Janeiro in 1992. Nicknamed the 'Earth Summit', its aim was to reconcile worldwide economic development with protection of the environment. It was big, with delegations from 178 nations, including 117 heads of state. Perhaps too big because the main treaty that emerged had no mandatory emission reduction targets. An attempt was made to put this right in 1997 via the Kyoto Protocol. It set a greenhouse gas reduction target of 5% by 2012. By 2012, instead of a 5% reduction there had been a 43% increase. GHG emissions had grown from 32 to 46 Gt/year (24 to 35 Gt/year CO_2, see Table 7.1).

As delegates gathered for the 21st Conference of the Parties (COP21) to the United Nations Framework Convention on Climate Change (UNFCCC) in Paris in December 2015, there was a sense of the last chance saloon. Urgency, determination, unity, collaboration and anxiety were there in spades and the result was possibly more than any had hoped for. After thirteen days there were 197 signatories to the Paris Agreement. The key elements of the agreement are:

- to limit global temperature increase to well below 2 degrees Celsius above pre-industrial levels.
- to pursue efforts to limit the temperature increase to 1.5 degrees Celsius above pre-industrial levels.
- to reduce greenhouse gas emissions to achieve a balance between anthropogenic emissions by sources and removals by sinks of greenhouse gases in the second half of this century (net zero emissions).
- for each country to prepare, communicate and maintain successive Nationally Determined Contributions (NDCs) that it intends to achieve.

The Paris Agreement represented a significant step forward in the global quest to address climate change, emphasising the importance of a collective and sustained effort to mitigate and adapt to its impacts. It is legally binding, although the NDCs determined by each country are not. The Agreement is designed to be flexible and inclusive, allowing for differentiated responsibilities and capacities among countries, while encouraging transparency and accountability.

Modelling helps us understand the intricacies of global warming and ocean acidification and evaluate the effectiveness of mitigation and adaptation strategies. By simulating climate processes, predicting future scenarios and assessing impacts, climate models provide a basis for informed decision-making and effective action. In this respect the IPCC reviews climate models from multiple sources to ensure a comprehensive and robust assessment of climate change. For the latest assessment report, AR6 issued in 2023, this involved over 100 models from the Coupled Model Intercomparison Project Phase 6 (CMIP6). The diversity and number of models reviewed help the IPCC provide a well-rounded understanding of climate dynamics, uncertainties and future projections.

This book looks into the future via a spreadsheet that analyses historic data sourced from the internet[33,34,35,36] and projects forward to the 2050 and 2100 milestones.

---

33 Ritchie, Hannah; and Roser, Max, 'Population Over the Long Run with Projections', Our World in Data, https://ourworldindata.org/grapher/population-long-run-with-projections?country=~OWID_WRL, accessed 18 December 2024.

34 Tans, Dr Pieter (NOAA/ESRL); and Keeling, Dr Ralph (Scripps Institution of Oceanography), Monthly mean CO_2 concentration, Mauna Loa 1958–2023, https://commons.wikimedia.org/wiki/Template:Other_versions/Mauna_Loa_CO2_monthly_mean_concentration, accessed 15 December 2023.

35 Ritchie, Hannah; and Roser, Max, 'Temperature Anomaly', Our World in Data, https://ourworldindata.org/grapher/temperature-anomaly, accessed 18 December 2024.

36 Ritchie, Hannah; and Roser, Max, 'CO_2 Emissions', Our World in Data, https://ourworldindata.org/co2-emissions, accessed 18 December 2024.

Since the greenhouse gas basket is effectively 76% CO_2, the model has been simplified by adopting CO_2 as the proxy for all greenhouse gases, and projections beyond 2023 are based on the following:

- 7.821 Gt of CO_2 released to the atmosphere will raise its atmospheric concentration by 1 ppm.
- The pre-industrial atmospheric concentration of 285 ppm has been adopted as a baseline. This equates to a baseline mass of 2.229 Tt of CO_2 in the atmosphere prior to 1850. At this time the global average temperature was between 13.7 and 14°C.
- In 1978 the global average temperature anomaly was 0.01°C and atmospheric CO_2 concentration was 335.42 ppm. In 2023 the global average temperature anomaly was 1.10°C and atmospheric CO_2 concentration was 421.08 ppm. During this period temperature increase was proportional to the growth of atmospheric CO_2 concentration at 0.0127°C/ppm. This factor has been adopted for calculating temperatures beyond 2023 and is equivalent to a rise of 3.6°C for a doubling of CO_2 concentration from 285 ppm.
- Atmospheric CO_2 in excess of the baseline CO_2 level of 2.229 Tt is considered 'excess atmospheric CO_2'. Every year a portion of this 'excess' is absorbed by land and sea. The amount absorbed annually is simply the difference between the annual emissions and the increase in residual CO_2. After 1950 this figure as a proportion of the excess CO_2 in the atmosphere has swung between one and three per cent. Between 1978 and 2023 the

spread was even less, so the average of 1.87% for this period has been adopted for projections beyond 2023.

- Estimates of how much CO_2 is absorbed by terrestrial ecosystems and oceans have been determined via direct measurements, modelling and synthesis of scientific research. The spreadsheet assumes the land/sea split for absorbed CO_2 is 50%, although this parameter could be varied to test sensitivity.

The process of projecting forward from 2023 begins with establishing a CO_2 emission target from 2024 to 2050. Global CO_2 emissions in 2023 were 37 Gt. By trial and error, it can be established that net zero is achieved by exponentially decaying 37 Gt CO_2 in 2023 to 23 Gt CO_2 in 2050. This results in a target of 36 Gt for 2024. Total CO_2 absorption for 2024 will be 1.87% of 2023 excess CO_2, that is, 19 Gt, leaving a 17 Gt increase in residual CO_2 for 2024, bringing the grand total of residual CO_2 in the atmosphere to 3.31 Tt. This is equivalent to 423 ppm CO_2 and this rise of 2 ppm of CO_2 from 2023 to 2024 will increase the global average temperature anomaly from 1.10 to 1.13°C. The process is then repeated for each year through to 2050 and ultimately 2100. Refer to Table 7.1.

Global Temperature Anomaly Projections beyond 2024 based on Annual Global Carbon Dioxide Emissions

Year	World Popn Bn	Atmos CO2 Conc ppm	ΔT oC	Residual CO2		Excess CO2 in Atmos Gt	CO2 Emiss'n Gt	CO2 Absorbed		Σ CO2 Ocean Absorb Gt
				in Atmos Tt	Incr Gt			Land & Ocean Gt	Rate %	
1850	1.3	285		2.23						
1860	1.4	286		2.24	2	11				
1870	1.4	288		2.25	2	21				
1880	1.5	291		2.27	5	45	1	-5		
1890	1.6	294		2.30	2	74	1	-0		
1910	1.8	300	-0.53	2.35	3	118	3	-0	-0.1%	
1920	1.9	303	-0.30	2.37	3	144	4	0	0.3%	4
1930	2.1	308	-0.18	2.40	2	176	4	2	0.9%	7
1940	2.3	311	0.08	2.43	1	206	5	4	2.0%	12
1950	2.5	311	-0.23	2.43	3	206	6	3	1.4%	37
1960	3.0	317	-0.12	2.48	7	251	9	2	1.0%	53
1970	3.7	326	-0.09	2.55	10	317	15	5	1.5%	78
1980	4.5	339	0.20	2.65	15	422	19	5	1.1%	114
1990	5.3	354	0.36	2.77	9	542	23	14	2.6%	157
2000	6.1	370	0.33	2.89	10	662	26	16	2.4%	217
2010	6.9	389	0.68	3.04	18	815	33	16	1.9%	288
2020	7.8	414	0.92	3.24	20	1,011	35	15	1.5%	369
2021	7.9	416	0.76	3.26	17	1,028	37	20	1.9%	378
2022	8.0	419	0.80	3.27	17	1,044	37	21	2.0%	389
2023	8.0	421	1.10	3.29	20	1,064	37	17	1.6%	397
2024	8.1	423	1.13	3.31	17	1,081	36	19	1.8%	407
2025	8.2	425	1.15	3.33	16	1,097	36	20	1.8%	417
2030	8.5	434	1.26	3.39	12	1,164	33	21	1.8%	468
2040	9.1	444	1.39	3.47	5	1,241	27	23	1.8%	579
2045	9.3	446	1.41	3.49	2	1,257	25	23	1.8%	636
2046	9.3	446	1.42	3.49	2	1,259	25	23	1.8%	647
2047	9.4	446	1.42	3.49	1	1,260	24	23	1.8%	659
2048	9.4	446	1.42	3.49	1	1,261	24	23	1.8%	670
2049	9.5	446	1.42	3.49	0	1,261	23	23	1.8%	682
2050	9.5	446	1.42	3.49	-0	1,261	23	23	1.8%	693
2051	9.6	446	1.42	3.49	-0	1,261	23	23	1.8%	705
2052	9.6	446	1.42	3.49	-0	1,261	23	23	1.8%	716
2053	9.7	446	1.42	3.49	-0	1,261	23	23	1.8%	728
2054	9.7	446	1.42	3.49	-0	1,261	23	23	1.8%	739
2055	9.8	446	1.42	3.49	-0	1,261	23	23	1.8%	751
2060	10.0	446	1.42	3.49	-0	1,260	23	23	1.8%	809
2070	10.3	446	1.42	3.49	-0	1,260	23	23	1.8%	924
2080	10.5	446	1.42	3.49	-0	1,259	23	23	1.8%	1,039
2090	10.5	446	1.42	3.49	-0	1,259	23	23	1.8%	1,154
2095	10.5	446	1.42	3.49	-0	1,259	23	23	1.8%	1,212
2096	10.4	446	1.42	3.49	-0	1,259	23	23	1.8%	1,223
2097	10.4	446	1.42	3.49	-0	1,259	23	23	1.8%	1,235
2098	10.4	446	1.42	3.49	-0	1,259	23	23	1.8%	1,246
2099	10.4	446	1.42	3.49	-0	1,259	23	23	1.8%	1,258
2100	10.4	446	1.42	3.49	-0	1,259	23	23	1.8%	1,269

Table 7.1 Climate data projections beyond 2024[37]

37 Global CO_2 emissions target for Net Zero 2050 is 23 Gt

The results are mixed. If global CO_2 emissions can be reduced to 23 Gt by 2050, then atmospheric CO_2 concentration will level off at 446 ppm and the global temperature anomaly will comfortably achieve the Paris Agreement goal of 1.5°C by peaking at 1.42°C. There will, however, be no further reduction in temperature below that level unless CO_2 emissions are reduced even further. Various scenarios are presented in Table 7.2. Even though it will be possible to reduce global temperatures by persevering with reductions, the CO_2 sequestered in the oceans will accumulate and ocean acidification will continue.

CO2 ppm	https://data.giss.nasa.gov/modelforce/ghgases/Fig1A.ext.txt		
CO2 Emissions	https://ourworldindata.org/co2-emissions		
Temperatures	https://ourworldindata.org/grapher/temperature-anomaly		
CO2 per ppm	7.821 billion tonnes		
Pre Industrial CO2	2.229 trillion tonnes		
Temp Coefficient	0.01	oC/ppm (1978 to 2023)	
Absorbed CO2 to Ocean	50%		
Excess Absorb Rate	1.83%	(1978 to 2023)	
2050 Target	23 Gt (billion tonnes)		Input Cell
2023 to 2050 Decay Constant	-0.018		
2100 Target	23 Gt (billion tonnes)		Input Cell
2050 to 2100 Decay Constant	0		

2050 Emission target scenarios										
		Atmos		Residual CO2		Excess		CO2 Absorbed		Σ CO2
	World	CO2		in		CO2 in	CO2	Land &		Ocean
	Popn	Conc	ΔT	Atmos	Incr	Atmos	Emiss'n	Ocean	Rate	Absorb
Year	Bn	ppm	oC	Tt	Gt	Gt	Gt	Gt	%	Gt
2023	8.0	421	1.10	3.29	20	1,064	37	17	1.6%	397
2050	9.5	446	1.42	3.49	-0	1,261	23	23	1.8%	693
2100	10.4	446	1.42	3.49	-0	1,259	23	23	1.8%	1,269
2023	8.0	421	1.10	3.29	20	1,064	37	17	1.6%	397
2050	9.5	446	1.42	3.49	-0	1,261	23	23	1.8%	693
2100	10.4	441	1.35	3.45	-1	1,220	21	22	1.8%	1,263
2023	8.0	421	1.10	3.29	20	1,064	37	17	1.6%	397
2050	9.5	446	1.42	3.49	-0	1,261	23	23	1.8%	693
2100	10.4	436	1.29	3.41	-3	1,179	19	22	1.8%	1,256
2023	8.0	421	1.10	3.29	20	1,064	37	17	1.6%	397
2050	9.5	446	1.42	3.49	-0	1,261	23	23	1.8%	693
2100	10.4	431	1.22	3.37	-4	1,138	17	21	1.8%	1,249
2023	8.0	421	1.10	3.29	20	1,064	37	17	1.6%	397
2050	9.5	443	1.37	3.46	-2	1,233	21	23	1.8%	691
2100	10.4	436	1.29	3.41	-1	1,182	21	22	1.8%	1,241
2023	8.0	421	1.10	3.29	20	1,064	37	17	1.6%	397
2050	9.5	443	1.37	3.46	-2	1,233	21	23	1.8%	691
2100	10.4	431	1.23	3.37	-2	1,142	19	21	1.8%	1,235
2023	8.0	421	1.10	3.29	20	1,064	37	17	1.6%	397
2050	9.5	443	1.37	3.46	-2	1,233	21	23	1.8%	691
2100	10.4	426	1.16	3.33	-3	1,102	17	20	1.8%	1,228
2023	8.0	421	1.10	3.29	20	1,064	37	17	1.6%	397
2050	9.5	439	1.33	3.43	-3	1,204	19	22	1.8%	688
2100	10.4	434	1.38	3.39	-1	1,165	21	22	1.9%	1,260
2023	8.0	421	1.10	3.29	20	1,064	37	17	1.6%	397
2050	9.5	439	1.33	3.43	-3	1,204	19	22	1.8%	688
2100	10.4	426	1.16	3.33	-1	1,104	19	20	1.8%	1,213
2023	8.0	421	1.10	3.29	20	1,064	37	17	1.6%	397
2050	9.5	435	1.28	3.40	-5	1,174	17	22	1.8%	685
2100	10.4	416	1.04	3.26	-2	1,026	17	19	1.8%	1,184
2023	8.0	421	1.10	3.29	20	1,064	37	17	1.6%	397
2050	9.5	435	1.28	3.40	-5	1,174	17	22	1.8%	685
2100	10.4	411	0.97	3.22	-3	987	15	18	1.8%	1,177

Table 7.2 2050 Emission target scenarios[38]

By 2050, assuming we have reduced global CO_2 emissions to 23 Gt, the oceans will have accumulated 693 Gt of CO_2 beyond

38 Global CO_2 emissions less than 23 Gt in 2050 will reduce Ocean CO2 absorption

pre-industrial levels. It is estimated pre-industrial oceans held 100 Tt of CO_2, so one could be forgiven for suggesting a further 0.693 Tt will be no more than the proverbial drop in the ocean. Unfortunately, that is not the case.

Sending pre-industrial CO_2 to the bottom of the ocean has kept phytoplankton, zooplankton, and mesopelagics busy for millions of years. All play crucial roles in the biological carbon pump, a process that transports sequestered CO_2 from the ocean surface to the seabed.

Phytoplankton are microscopic marine algae that live in the sunlit upper layers of the ocean (the euphotic zone). The global biomass of phytoplankton is estimated to be around 1 to 2 Gt of carbon. They use sunlight to perform photosynthesis, converting dissolved CO_2 into organic matter (biomass) and releasing oxygen. This is the foundation of the oceanic food web. Through photosynthesis, phytoplankton fix large amounts of carbon. This carbon is then passed on to other organisms as they are consumed.

Zooplankton are small, often microscopic, animals that feed on phytoplankton. As they consume phytoplankton, they incorporate carbon into their own biomass. The global biomass of zooplankton is estimated to be approximately 0.2 to 0.8 Gt of carbon. They produce faecal pellets that are rich in carbon and, being dense, sink rapidly, carrying neutralised carbon from the surface to deeper ocean layers. Some zooplankton perform daily vertical migrations, moving to deeper waters during the day to avoid predators and rising to the surface at night to feed. This migration is another means of transporting carbon vertically in the ocean.

Then there are the mesopelagic fish, such as lanternfish, that feed on zooplankton. They live in the mesopelagic zone of the ocean, typically between 200 and 1000 metres deep, and with an estimated biomass of between 1 and 10 Gt, comprise one of the most abundant fish groups in the ocean. They perform daily vertical migrations similar to zooplankton. They ascend to the surface at night to feed on zooplankton and other prey, then descend to deeper waters during the day. This migration is one of the largest movements of biomass on Earth and plays a significant role in transporting carbon to the deep ocean. Like zooplankton, mesopelagic fish contribute to carbon sequestration through the production of faecal pellets, which sink to the seabed.

The combined activities of these organisms facilitate the transfer of carbon from the surface to the ocean depths, where it can be sequestered in sediments, playing a key role in the global carbon cycle and helping to mitigate atmospheric CO_2 levels. But unfortunately, they cannot work fast enough to keep up with the extra quantities of CO_2 that have been absorbed by the oceans since 1912. Table 7.1 indicates that between 1912 and 2023, 397 Gt of CO_2 were absorbed by the world's oceans. As already alluded to, this has reduced background oceanic pH from 8.2 to 8.05. This doesn't seem like much, but the scale is logarithmic and indicates that oceanic alkalinity has been reduced by 30%.

Assuming global emissions can be reduced to 23 Gt of CO_2 and net zero is achieved by 2050, an estimated 693 Gt of additional CO_2 will still have been absorbed by the oceans since 1912 and background pH will have dropped to around 7.95, thus reducing ocean alkalinity even further. This prospect has given climate scientists and marine biologists further cause for concern.

Chapter 8

Ocean Acidification and the Great Barrier Reef

Ocean acidification in the post-industrial era poses a significant threat to marine ecosystems and global environmental health, and there is the potential for irreversible long-term effects.

Many marine organisms, such as corals, molluscs and certain plankton species, rely on calcium carbonate ($CaCO_3$) to form their shells and skeletons. As ocean pH decreases (becoming more acidic), the availability of carbonate ions, which are essential for calcium carbonate formation, also decreases (see equations p 32). This makes it more difficult for these organisms to maintain and build their structures. Early laboratory studies concluded coral reefs were particularly vulnerable. This made sense since acidification should weaken coral skeletons, making them more susceptible to erosion and less able to recover from bleaching events. Since healthy coral reefs provide habitat and food for a vast array of marine species, they are critical for biodiversity. Organisms like oysters, clams and mussels would also struggle to build their shells, leading to reduced survival rates and the obvious economic impact on fisheries and aquaculture.

Plankton, the base of many marine food webs, are also thought to be affected by acidification. Some planktonic species, like pteropods (sea butterflies), have calcium carbonate shells that are particularly sensitive to pH changes. Any decline in these species would disrupt food webs, impacting species that rely on them as a primary food source, including fish, birds and marine mammals.

Fortunately, it's not as bad as it sounds. Recent research has indeed shown that some marine organisms, particularly those that construct calcareous structures (such as shells and skeletons), may be more resilient to ocean acidification than previously thought. While ocean acidification remains a significant threat, these findings suggest that certain species have adaptive mechanisms or inherent robustness that allow them to cope with changing ocean chemistry.

Some species have shown a remarkable ability to adapt to lower pH levels through natural selection. Populations that live in areas with naturally fluctuating pH levels, such as coastal upwelling zones, appear to have developed genetic traits that make them more resistant to acidification.

Certain marine organisms exhibit phenotypic plasticity, meaning they can adjust their physiology, morphology or behaviour in response to environmental changes. For example, some species of corals and molluscs can modify their calcification processes to maintain their shells or skeletons even in more acidic conditions.

Further research has shown that some organisms can regulate their internal pH and continue to produce calcium carbonate structures even when external conditions become more acidic. This ability to maintain calcification rates is crucial for their survival. Other organisms produce organic layers or coatings around their calcareous structures, which can protect against acidification by providing a barrier between the skeleton or shell and the surrounding sea water.

A well-known example of this is the Great Barrier Reef (GBR) off the east coast of Queensland. It is one of the seven natural wonders of the world and the largest coral reef system on the planet. Its sheer size, beauty and biodiversity make it a national icon, representing the extraordinary natural heritage of Australia. It holds a special and significant place in the Australian psyche, symbolising not just natural beauty and biodiversity but also national pride, cultural identity and environmental stewardship. It is also a major tourist attraction, drawing millions of visitors from around the world. It supports a multi-billion-dollar tourism industry, which is a vital part of the economies of Queensland and Australia as a whole.

The reef is often seen as a symbol of Australian identity, embodying the country's connection to the ocean, love of the outdoors and reputation as a land of extraordinary natural landscapes; so it has become central to discussions about conservation and environmental protection. It is also a major element in the national debate over climate change. GBR health is often cited in discussions about Australia's environmental policies, fossil fuel use and global climate commitments. Much of Queensland's coal is exported in giant bulk carriers that have no option but to negotiate passages through the reef. From time to time, one runs aground and, in many ways, does the reef a favour by reigniting debate over balancing economic growth with environmental sustainability.

This debate is never far from the academic common rooms and classrooms of Townsville's James Cook University (JCU), which is closely connected to the reef through its location, research and educational programs. It had a reputation as one of the world's leading centres for marine science until 2018 when one of its senior researchers, Dr Peter Ridd, publicly criticised research conducted by his colleagues regarding the reef's health. He argued some of the research was flawed, exaggerated or misleading, particularly studies claiming the reef was in significant danger from climate change and other factors. In 2018, JCU terminated Ridd's employment, citing repeated breaches of the university's code of conduct. What wasn't cited were the pots of research funding put at risk by Peter Ridd's revelations. Ridd subsequently sued JCU, arguing his termination was unlawful and violated his right to academic freedom. He lost the legal battle and his job, but the Great Barrier Reef carries on regardless, just like 'Ol' Man River'.

The Australian Institute of Marine Science (AIMS) has been monitoring the reef for thirty-eight years. It's a massive undertaking. The 2900 reefs and 900 islands of the GBR stretch over 2300 kilometres and cover an area of 344 400 square kilometres. The reef system is so vast it can be seen from space. Unfortunately, it can't be monitored from space. AIMS surveys the reef with divers towed by motorboats. The primary objective of the surveys is to measure per cent hard coral cover, which has been adopted worldwide as a practical indicator of reef health.

The reef experienced its lowest recorded hard coral cover in 2017. This decline was largely caused by a combination of extreme disturbances, including back-to-back mass coral bleaching events in 2016 and 2017, as well as severe tropical cyclones and crown-of-thorns starfish outbreaks. During this period, the average coral cover on the GBR fell significantly, reaching a level of 13% over the northern GBR in 2017 and 14% over the central GBR in 2019. The last report on GBR condition was issued on 7 August 2024.[39] With hard coral cover at 39%, 34% and 39% on the northern, central and southern reefs it's been the best year yet. It's unfortunate the report was not issued in time to save Peter Ridd.

None of this is to say climate change isn't a threat. We may be able to arrest global warming by reducing fossil fuel consumption, but relying on the oceans to digest accumulating CO_2 is untenable. The AIMS surveys show that GBR corals are resilient to increasing ocean acidification for the time being, but this can't last forever. While there is still time, we need to eliminate fossil fuels for everything but the most essential uses. Only then, and after disposing of the GHG equivalent by methods such as direct air capture, will we be able to rest easy. In other words, real zero not net zero.

Australia prides itself on being a smart country. And it's a rich one as well. If anybody can do this, the Lucky Country can ... perhaps.

39 *AIMS Long-Term Monitoring Program, Great Barrier Reef Annual Summary Report, Coral Reef Condition 2023–2024*, Australian Institute of Marine Science, 7 August 2024, https://www.aims.gov.au, accessed 1 November 2024.

Chapter 9

The Lucky Country's Nightmare

John Howard (Coalition 1996–2007) was generally sceptical about the urgency of climate change. His government was reluctant to commit to binding emissions targets and did not ratify the Kyoto Protocol. Towards the end of his tenure, however, and possibly in response to *An Inconvenient Truth*, Howard recognised growing concern and proposed a national emissions trading scheme. But it was too late to save the 2007 general election. Climate change was an important issue and his government's reluctance to take decisive action, coupled with Kevin Rudd's strong commitment to environmental policies, contributed to a broad perception that Howard's leadership was out of step with the times. While economic and industrial relations issues also played significant roles, climate change was a defining issue in Howard's loss, especially among younger and environmentally conscious voters. He lost the general election and his seat, perhaps his only consolation being he wasn't betrayed by his colleagues. But he was scalp #1 nevertheless.

Kevin '07' Rudd (Labor 2007–2010, 2013) was proactive and committed and positioned himself as a strong advocate for action on climate change. One of his first acts as prime minister was to ratify the Kyoto Protocol, signalling Australia's commitment to global climate efforts. He also proposed the Carbon Pollution Reduction Scheme (CPRS), a cap-and-trade system, but the opposition leader at the time, Tony Abbott, tore the idea to shreds, and when it became clear Rudd would lose the 2010 general election, Julia Gillard (Labor 2010–2013) challenged him to a leadership ballot. He resigned, and in June Gillard took over as prime minister, unopposed. Scalp #2.

Julia called an election in August, stridently declaring during a televised campaign event, 'There will be no carbon tax under the government I lead'. She won the election but not with a majority so was forced to form a minority government with a ragtag bunch of independents and a seething Kevin Rudd. He was appointed foreign minister, presumably to keep him as far away as possible. It was never going to end well. She soldiered on with climate action, including a carbon pricing mechanism incorporated in a broader Clean Energy Future package. Tony Abbott dubbed it a 'carbon tax', claimed it was an economic 'wrecking ball' and never missed an opportunity to bash her with it. Three years later when it was clear she couldn't win the 2013 election, Kevin Rudd saw his opportunity for revenge, rallied the numbers for a leadership ballot and won. It was a pyrrhic victory. Three months later he lost the general election and Tony Abbott (Coalition 2013–2015) became Australia's third prime minister in six years. Julia may have been scalp #3, but the socialists weren't the only backstabbers in Canberra.

When John Howard lost the general election and his seat in November 2007, Brendan Nelson, former defence minister, became Liberal Party leader and leader of the Opposition, narrowly defeating the upstart Malcolm Turnbull in a party room vote. His victory was seen as a victory for the Liberal Party's conservative faction. Turnbull represented the more moderate, progressive faction. These divisions were never seen under John Howard but surfaced soon after his departure, and climate change was the catalyst.

Nelson faced several challenges. The Liberal Party was still grappling with the fallout from its election defeat, and he struggled to gain public traction and support from within the party. The major issue was the party's stance on climate change. Nelson initially supported Rudd's plan for an emissions trading scheme but later withdrew it, sowing confusion and dissatisfaction within the party and among the public. The Liberal Party was hopelessly divided over climate change, with moderates pushing for stronger action and conservatives sceptical of such measures. Nelson's wavering on the issue frustrated both factions, and in September 2008 Turnbull pounced, forcing a leadership ballot, which he won to become the new Opposition leader.

By then, Kevin Rudd's focus as prime minister was the implementation of the CPRS and managing the impacts of the global financial crisis. Turnbull decided to support the CPRS without sufficiently consulting his conservative faction. This created a rift, and in December 2009 Tony Abbott forced a leadership ballot, won it by one vote and became the new Opposition leader. Abbott proved to be a ruthless leader and was undoubtedly the catalyst for the downfall of both Kevin Rudd and Julia Gillard.

Upon becoming prime minister in 2013, Abbott moved quickly to repeal Gillard's carbon pricing mechanism. His approach was focused on reducing costs for businesses and consumers rather than pursuing aggressive climate policies. He favoured a 'direct action' plan, which included an emissions reduction fund as an alternative to carbon pricing. These initiatives faced strong opposition from environmental groups, scientists and a growing segment of the public but were nothing compared to the widespread ridicule and bewilderment, both within his own party and among the electorate, that followed his award of an Australia Day knighthood to Prince Philip.

As minister for communications and architect of the deeply flawed National Broadband Network, Malcolm Turnbull (Coalition 2015–2018) had been stalking his prime minister for months. The Prince Philip knighthood fiasco gave him the justification he needed to launch a leadership ballot, which he won unopposed. In Opposition, Turnbull had been an advocate for strong climate action. His approach as prime minister, however, was constrained by the need to maintain party unity, particularly with the conservative faction. Turnbull supported the Paris Agreement and proposed the National Energy Guarantee (NEG), which sought to balance emissions reductions with energy reliability and affordability. However, internal party divisions limited his ability to implement more robust climate policies. In August 2018 he was ousted in a leadership ballot and became the fifth Australian prime minister to fall foul of climate change policy. Scalps #4 and #5.

The winner from nowhere was Turnbull's treasurer, Scott Morrison (Coalition 2018–2022). His approach to climate change was pragmatic, focusing on energy security and economic impacts. He supported Australia's commitments under the Paris Agreement but was criticised for lacking ambition in climate policy. Morrison emphasised the importance of fossil fuels, particularly coal, in Australia's energy mix. His government promoted the use of gas as a transition fuel and invested in carbon capture and storage technologies. However, Morrison faced criticism for what many saw as insufficient action on climate change, particularly during the 2019–2020 bushfire crisis. He might have survived the bushfire imbroglio but was unable to weather the Brittany Higgins allegations that she was sexually assaulted by a male colleague in a ministerial office within Parliament House in Canberra. The affair prompted a major political and public outcry, leading to widespread discussions about the treatment of women in politics, workplace culture and issues of sexual harassment and assault within Parliament. The Coalition didn't stand a chance in the general election that followed shortly afterwards, and Mr Morrison lost his job. Certainly, climate change was a factor, but he couldn't be described as a climate change scalp. Anthony Albanese (Labor 2022 –present) became Australia's latest prime minister in May 2022.

Without doubt Albanese's government has gone a long way towards restoring Australia's reputation in international climate forums. The new government immediately revised

Australia's Nationally Determined Contribution, accelerating the country's commitment to reducing GHG emissions by moving the interim 2030 target from 28% to 43% below 2005 levels, with net zero 2050 unchanged. The plan focuses on several key strategies and initiatives. It aims to balance economic growth, energy security and emissions reduction by leveraging technology, innovation and policy frameworks.

The key technologies include expanding the use of solar, wind and other renewable energy sources along with storage solutions like batteries and pumped hydro; lifting carbon storage in agricultural soils through improved sequestration practices; developing and deploying carbon-free steel and aluminium manufacturing processes; encouraging carbon capture and underground storage for big industrial carbon emitters; and developing Australia as a major exporter of clean hydrogen.

The government's plan for the transformation of Australia's power generation and distribution network has been prepared by the Australian Energy Market Operator (AEMO) and is presented in its 2024 Integrated System Plan for the eastern states. According to the plan, 'The lowest-cost way to supply electricity to homes and businesses throughout Australia's transition to a net zero economy' is 'renewable energy connected by transmission, firmed with storage and backed up by gas-fired generation'.[40]

So far so good, but the plan shows little has been learned from the divisions of the past twenty years. The decision to go 'renewables' has been taken without giving nuclear power

40 Australian Energy Market Operator (AEMO), 2024 Integrated System Plan for the National Electricity Market – A Roadmap for the Energy Transition, June 2024, https://aemo.com.au.

more than a mention in passing and appeared as more of a political rather than rational decision. Apart from antipathy to nuclear within its own ranks, Labor relies heavily on Greens Party preferences to win general elections. So, for the Albanese government, nuclear was not an option and the excuse was cost.

Before going further, it's useful to return to the good old days when there was no need to call your accountant to decipher your power bill or your banker to pay it. The power came 24 hours a day, 365 days a year from thumping heart coal-fired power stations like Liddell, Mount Piper, Loy Yang, Tarong and Callide, and they delivered electricity at the farm gate for around A$60/MWh (A$0.06/kWh) in 2024 dollars. This was sold on to domestic consumers for around A$0.20/kWh, in 2024 dollars.

Fast forward to real 2024 and it's a different story. Of the 80 GW of currently installed capacity, 32 GW is coal and gas. The rest, 48 GW, is generated from renewables. The Australian Labor Government never talks about 'renewable energy'; it's always 'cheap renewable energy', so it would be reasonable to assume domestic consumers now pay less than A$0.20/kWh for their electricity. Not so. Domestic consumers now pay A$0.35/kWh. The government pretends it's not happening, but it is, and as the contribution from renewables grows it's likely to get worse. Maybe this is the price to be paid to arrest global warming.

The Opposition doesn't think so. They look to France and Germany who provide a real comparison between nuclear and renewables. Seventy per cent of France's power is nuclear. With just one fatality thirteen years ago, it's safe and reliable as well. Just to be different, Germany is phasing out nuclear and has opted for renewable power, which now provides around half its demand.

The levelised cost of electricity (LCOE) for nuclear power in France is €40 to €60 per MWh, which is relatively low due to the long operational life of reactors and their ability to provide steady baseload power. This gives France an advantage in terms of cost-effective electricity production compared to countries relying more on fossil fuels or renewables with higher integration costs.

Although the LCOE from wind and solar in Germany is similar to that of nuclear in France, the integration costs for wind and solar (due to intermittency and the need for backup power and grid upgrades) push up the LCOE. Then there is the continued use of coal and natural gas, which attract a European carbon tax and raise the overall LCOE in Germany to €75 per MWh. The sad irony is that if the €500–600 billion Germany has spent on renewables over the past two decades to provide 50% of the power it needs had been spent on nuclear energy, it would now be generating all the power it needs from clean zero emission sources. Could it be a similar story in Australia?

In June 2024, Peter Dutton, the Australian Opposition leader, announced that should the Coalition win the 2025 general election they would reduce electricity costs by building seven nuclear plants of unspecified capacity on sites made vacant following retirement of the coal-fired power stations mentioned earlier. The big idea was to save cost by utilising the existing transmission lines linking each of those sites to the grid. These costs are substantial at over $7 million per kilometre. There was also a suggestion that NuScale small modular reactors (SMRs) would be involved. The plan was cheered by many but took a cold bath shortly afterwards when the first commercial deployment of NuScale SMR technology in the USA was shelved due to cost overruns. Mr Albanese was delighted.

There is insufficient information in Mr Dutton's nuclear proposal to support an objective evaluation; however, it is possible to compare the capital cost of AEMO's 'all renewables' plan to be achieved by 2050 with an equivalent 'all nuclear' plan. Most of the cost data required to undertake this study is provided in a report prepared by Australia's national science agency, CSIRO, entitled 'GenCost 2023–24'. One key element not provided, however, is the cost of transmission lines. No wonder; no two are the same.

According to AEMO's 2024 Integrated System Plan, the eastern states will require 'around 5000 km of transmission to be delivered over the next decade', with a further 5000 km by 2050. The plan identifies fifteen projects (see Table 9.1), for a total of 4937 km of transmission lines at a cost of $35 billion, that is, $7100/km, suggesting that by 2050 the federal and state governments will have spent $71 billion installing the transmission lines necessary to 'rewire the nation' for renewables.

Network projects in the 2024 ISP optimal development path							
Project Title	From	To	Length km	Load MW	Spec kV	Budget $M	Online
Far North Qld REZ	Woree (Cairns)	Ross			275		2024
EnergyConnect	Robertstown	Wagga Wagga	900	800	330	$1,800	2027
Western Renewables Link	Bulgana	Sydenham	190		500	$370	2027
Central West Orana REZ TL	Elong Elong	Wollar	90	4,500	500	$1,600	2028
Central West Orana REZ TL	Gathering	Lines	150		330	$1,600	2028
Copperstring 2032	Mt Isa	Cloncurry	110	6,000	220	$1,000	2029
Copperstring 2032	Cloncurry	Hughenden	400	6,000	230	$2,000	2029
Copperstring 2032	Hughenden	Townsville	340	6,000	500	$2,000	2029
HumeLink	Wagga Wagga	Bannaby	365	2,200	500	$4,900	2026
Sydney Ring	Bannaby	Sydney	114		500	$1,600	2031
Hunter Transmission Project	Olney/Eraring	Bayswater	110		500	$1,100	2028
Gladstone Grid Reinforcement	Callide	Gladstone	90		275	$1,300	2029
Mid NS Australia REZ Expansion	Bundey	Parafield Gdns	132		275	$389	2029
Waddamana to Palmerstone	Waddamana	Palmerston				$0	2029
VNI West	Jerilderie	Bulgana	500		500	$3,900	2028
Marinus Link Stage 1	Burnie	Latrobe Valley	345	750		$3,800	2030
Marinus Link Stage 2	Burnie	Latrobe Valley	345	750		$2,700	2032
New England REZ TL	Armidale	Bayswater	560	3,000		$1,800	2033
Queensland SuperGrid South	Borumba	Halys	106		500	$1,800	2031
Queensland SuperGrid North	Borumba	Woolooga	90		500	$1,500	2031
			4937			$35,159	
						$7	
AEMO Network Projects from 2024 Integrated System Plan page 14.							

Table 9.1 Rewiring the Nation

The backup required for renewable power will be provided by 15 GW of gas-fired power. The open-cycle gas turbine (OCGT) is the most suitable technology for gas-powered generation with frequent start-up and shut-down requirements. It provides rapid response and flexibility, making it ideal for peaking plants or backup generation when quick power is needed to stabilise the grid or respond to sudden demand fluctuations.

The basic elements of AEMO's plan to transition to net zero by 2050 with 'renewable energy connected by transmission, firmed with storage and backed up by gas-fired generation' are tabled below (Table 9.2) together with the author's estimate to provide the same quantity of dispatchable power with zero-emission nuclear power generation plants. The calculations assume that 86 GW of rooftop solar and 7 GW of hydro power are common to both scenarios and that the renewables and nuclear scenarios will require ten and five thousand kilometres of transmission lines, respectively.

If the average capacity factors and capex costs published in the GenCost 2023–24 report[41] are applied, AEMO's planned 220 GW of renewable capacity will provide 66 GW of dispatchable power at a cost of $482 billion. This is 9 GW short of the 75 GW called for in the AEMO plan but still sufficient to supply the anticipated annual demand of 410 TWh called for in the same plan.

On the basis outlined on the next page, the renewables capex at $482 billion is 19% less than the nuclear capex at $573 billion. These numbers don't include operating costs, but since the capex gap is reasonably wide and since there are

41 Graham, P.; Hayward, J.; and Foster, J, GenCost 2023–24: Final Report, CSIRO, 2024.

negligible fuel costs it would not be unreasonable to assume the renewables LCOE will be less than that for nuclear. So, based on this analysis renewables should be the preferred option and, burgeoning electricity costs aside, so far so good. But that's because there is still enough coal-fired capacity to cover for the problems that are already surfacing as new wind and solar projects are commissioned. This will only get worse as coal-fired cover recedes and the point of no return comes and goes.

That's why we need a Plan B.

Renewables capex vs nuclear capex										
Network Element	Unit	Unit Cost $Bn	AEMO Renewables 2050				Nuclear 2050			
			Rated Capacity GW	Av Cap Factor	Dispatch GW	Capex $Bn	Rated Capacity GW	Av Cap Factor	Dispatch GW	Capex $Bn
Rooftop PV*	/GW		86	0.17	15		86	0.17	15	
Utility Solar	/GW	$1.45	58	0.26	15	$84		0.26		
Onshore Wind	/GW	$2.94	62	0.39	24	$182		0.39		
Offshore Wind	/GW	$5.50	7	0.46	3	$38		0.46		
Flexible Gas	/GW	$1.02	15	0.20	3	$15		0.20		
Hydro*	/GW		7	0.95	7		7	0.95	7	
Nuclear	/GW	$8.45		0.71			64	0.71	45	$538
Utility Storage	/GW	$5.00	18			$90				
Transmission	/km	$0.01	10,000			$71	5,000			$36
					66	$482			66	$573
*Rooftop PV and Hydro common to both scenarios and excluded from cost comparison AEMO Renewables Data from 2024 Integrated System Plan pages 11, 12 and 47. Capacity Factors from CSIRO GenCost 2023-2024 Final page 91. Capex Estimates from CSIRO GenCost 2023-2024 Final page 82										

Table 9.2 Renewables capex vs nuclear capex

Chapter 10

Plan B

Antipathy to nuclear power generation in Australia goes back to the British nuclear tests in South Australia in the 1950s and 1960s. They left a legacy of health and environmental issues and a public mistrust of nuclear technology in general. This was exacerbated during the 1970s and 1980s by concerns over nuclear waste, nuclear accidents (Three Mile Island, Chernobyl) and environmental impacts. As a result, the *Environment Protection and Biodiversity Conservation Act* (EPBC Act) was passed in 1999. It included provisions effectively banning the construction of new nuclear power facilities in Australia except for the Lucas Heights reactor used for scientific and medical research.

As discussed in previous chapters, the Australian Government's plan to replace eastern Australia's coal-fired power generation facilities with renewables by 2050 is presented in AEMO's 2024 Integrated System Plan. The novelty and complexity of the undertaking is underscored, and a chapter devoted to the anticipated risks talks about infrastructure investment delays, market and policy setting delays and loss of social licence. But they didn't want to frighten the horses so made no mention of the inevitability of cloudy days with no wind. Denmark and Ireland, the Europeans with the most advanced renewable energy systems, both rely heavily on backup power from neighbours.

Denmark plans a 100% renewable energy system by 2050. Currently it has 6.7 GW of wind power and 1.8 GW of solar backed up by 2.4 GW of natural gas, 1.6 GW of biomass and 6 GW of grid interconnection with neighbours Norway, Sweden and Germany. That's 118% firm power backup for the renewables.

Ireland is transitioning rapidly towards an 80% renewable energy power generation mix, with a strong emphasis on wind power. But Ireland also relies on a combination of natural gas and interconnections with neighbouring countries to provide backup during periods of calm. It is already a long way down the track with 5.0 GW of wind backed up by 4.7 GW of natural gas and 1.5 GW of grid interconnection with the UK. That's 124% firm power backup.

Ireland also has plans for the 'Celtic Interconnector', a €1 billion, 500 km high-voltage direct-current submarine cable between La Martyre in France and Cork. This will provide further backup for its renewables and, believe it or not, lead to more competitive electricity prices by giving Ireland access to France's lower-cost nuclear power.

As can be derived from Table 9.2, AEMO's plan for net zero 2050 is for 213 GW of wind and solar backed up by 22 GW of gas and hydro – that's 10% firm power backup, and there is no big brother next door to come to the rescue when things go wrong.

And they do. Transmission losses have been underestimated, capacity factors have been overestimated, and there is already output curtailment around noon each day to avoid negative wholesale prices.

The conversion from 'Rated Capacity' to 'Dispatch' (dispatchable power) in Table 9.2 adopts the mid-range capacity factors tabled in CSIRO's GenCost 2023–24 report. Accordingly, 62 GW of onshore wind, with a capacity factor of 39%, converts to 24 GW of dispatchable power. But these numbers ignore transmission losses ranging from 5 to 20% for remote projects. As a result, ongoing transmission infrastructure upgrades are critical.

Of the 62 GW of onshore wind planned for 2050, 13.3 GW has been installed to date. Since 2011 the capacity factor for the installed assets has been hovering around 30%,[42] not the 39% mentioned in Table 9.2. This means the renewables network as planned is not just 9 GW but 14 GW short of the required 75 GW.

Then there is the 'curtailment' problem. Renewable Energy Zones (REZs) in Australia's eastern states are designed to coordinate the development of large-scale renewable energy projects (like solar, wind and storage) with the required grid infrastructure. These zones will help concentrate renewable energy generation in areas with high-quality wind and solar resources, allowing for efficient transmission to major demand centres.

It sounds like a good idea, but on good days, particularly for solar farms, the operators have problems getting their

42 Mcardle, Paul, An updated look at monthly NEM-wide wind statistics, Watt Clarity, https://wattclarity.com.au/articles/2024/09/08sept-monthly-wind-statistics/, accessed 22 October 2024.

power to market through overloaded transmission lines and when they do there is no market for their power anyway. So, they shut down and these shutdowns can last for weeks. From the investors' point of view, it's revenue that can never be recovered unless they install batteries, which are expensive.

The following table (Table 10.1) recalculates the dispatchable power available from the AEMO renewables plan adopting revised capacity factors considering the issues raised above. This reduces the dispatchable power available from the renewables plan from 66 to 57 GW for the same cost as before, that is, $482 billion. This will produce 498 TWh in 2050, which will satisfy the anticipated demand of 410 TWh.

Renewables capex vs revised nuclear capex										
Network Element	Unit	Unit Cost $Bn	AEMO Renewables 2050				Nuclear 2050			
			Rated Capacity GW	Reduced Capacity Factors	Dis-patch GW	Capex $Bn	Rated Capacity GW	Av Cap Factor	Dis-patch GW	Capex $Bn
Rooftop PV*	/GW		86	0.17	15		86	0.17	15	
Utility Solar	/GW	$1.45	58	0.23	13	$84		0.26		
Onshore Wind	/GW	$2.94	62	0.27	17	$182		0.39		
Offshore Wind	/GW	$5.50	7	0.36	3	$38		0.46		
Flexible Gas	/GW	$1.02	15	0.20	3	$15		0.20		
Hydro*	/GW		7	0.95	7		7	0.95	7	
Nuclear	/GW	$8.45		0.71			50	0.71	36	$425
Utility Storage	/GW	$5.00	18			$90				
Transmission	/km	$0.01	10,000			$71	5,000			$36
					57	$482			57	$461

*Rooftop PV and Hydro common to both scenarios and excluded from cost comparison

AEMO Renewables Data from 2024 Integrated System Plan pages 11, 12 and 47.

Capacity Factors reduced to account for transmission losses, reduced wind farm performance and "curtailment".

Capex Estimates from CSIRO GenCost 2023-2024 Final page 82

Table 10.1 Renewables capex versus revised nuclear capex

To match this with nuclear power requires just 50 GW with an all-in cost of $461 billion (Table 10.1). Hey presto, nuclear becomes the cheaper option obviating the necessity for 15 GW of 'flexible gas', thereby saving five million tonnes of CO_2 annually.

Before exploring how nuclear power generation might be introduced into Australia's power generation mix, a brief review of the industry will be useful. The global nuclear industry is currently undergoing a period of both expansion and challenge. Many countries are continuing to invest in nuclear power as part of their efforts to transition to low-carbon energy systems, but the high costs of nuclear plants and lengthy construction times remain significant barriers.

Despite this there are currently 440 reactors with a capacity of 394 GW operating in thirty-two countries producing 10% of the world's electricity. This accounts for 25% of the world's low-carbon electricity, contributing significantly to climate change mitigation efforts. As the war on carbon intensifies so does the demand for nuclear reactors. There are currently fifty-nine under construction worldwide to add an estimated 60 GW of nuclear capacity when completed. China is the leader in this crusade with twenty-four reactors (~25 GW) under construction and plans for a further 150 GW by 2060. France is not far behind with a plan to build up to fourteen new reactors by 2050 to replace ageing plants.[43]

But it's not all plain sailing. High-profile accidents such as Chernobyl in 1986 and Fukushima in 2011 have stirred public and political resistance, particularly in countries like Germany

43 World Nuclear Performance Report 2024, World Nuclear Association, London, https://world-nuclear.org, accessed 5 November, 2024.

and Japan, not to mention Australia. On top of that, nuclear power is an expensive form of electricity generation due to long construction periods (ten plus years for large reactors) and complex safety and regulatory frameworks. This is why there has been a trend away from the big projects to the concept of small modular reactors.

SMRs represent a new generation of nuclear reactors designed to provide scalable, flexible and potentially more affordable nuclear power. SMRs are defined as reactors with a capacity of up to 300 MW per unit, significantly smaller than traditional nuclear reactors, which typically generate between 1 and 1.6 GW. SMRs are designed to be produced and tested in factories, transported to and assembled on-site at a lower cost than stick building on-site.

The key feature of an SMR is that it is fabricated and tested by an experienced workforce in the controlled environment of a factory. This reduces on-site construction complexity and time. They also incorporate advanced safety designs, such as passive cooling systems, which allow the reactors to shut down safely in the event of an emergency, without external power or human intervention.

The world's most advanced SMR is China's Linglong One project. An extract from a China Atomic Energy Authority press release dated 10 August 2023, reads as follows.

> On August 10th, the core module of the Linglong One reactor, a multi-purpose modular small reactor in Hainan, a subsidiary of China National Nuclear Corporation (CNNC), was lifted and moved towards the nuclear island. Pressure vessel, evaporator

> and other key equipment were installed in one step, marking the peak phase of Linglong One's installation.
>
> The Hainan Nuclear Power Multi-purpose Modular Small Reactor Technology Demonstration Project is the world's first commercial land-based small modular pressurized water reactor. Upon its operation, it is expected to generate one billion kilowatt-hours of electricity per year, which could meet the needs of 526 000 households and is equivalent to reducing around 880 000 tons of CO_2 emissions or planting 7.5 million trees. With its stable and reliable clean energy supply, it will contribute to China's carbon peaking and carbon neutrality goals and the construction of a 'clean energy island' in Hainan, and drive the development of nuclear energy worldwide.[44]

Also known as the ACP100 SMR, Linglong One is a major milestone in China's drive to develop advanced nuclear technology. It's been developed by the China National Nuclear Corporation and represents China's first domestically designed SMR.

China has never been held back by health, safety or environmental regulations, so the project is still on schedule. The 125 MW design incorporating passive safety features such as natural circulation for cooling and passive containment cooling systems was approved by the International Atomic Energy Agency in 2016. Construction at the Changjiang site

44 'World's first! The Heart of Linglong One is in place', China Atomic Energy Authority, Press Release, 10 August 2023, https://www.caea.gov.cn/english/n6759361/n6759362/c10087720/content.html, accessed 20 December 2024.

in Hainan Province began in July 2021, and the reactor is due for commissioning in 2026. Once operational, it will be the world's first land-based commercial SMR.

Figure 10.1 Installing the Linglong One reactor at Changjiang (Photo: press release, footnote 44)

Another advanced design is the RITM-200 developed by Rosatom, the Russian state corporation responsible for the country's nuclear energy industry. Rosatom was established in 2007 and operates across the entire nuclear value chain, from uranium mining and enrichment to reactor design, construction and decommissioning. It is involved in both the civilian nuclear power sector and nuclear research, as well as providing nuclear technology exports. RITM-200s have been installed in Russia's latest Arktika-class nuclear icebreakers operating in the Arctic but can also be used for onshore power generation.

Figure 10.2 Arktika-class nuclear icebreaker at work[45]

The RITM-200 SMR is a pressurised water reactor fuelled by enriched uranium. It comes in 55 MW modules, which can be installed as multiple units. Rosatom is promoting the SMR both domestically and internationally for civilian power generation in remote areas. The first onshore RITM-200 SMR is earmarked for deployment in Yakutia by the early 2030s.

The competition is still a way behind China and Russia; however, there are several designs in the pipeline, by companies including NuScale Power (USA) and Rolls-Royce (UK). NuScale is sponsored by Fluor Corporation, a major US engineering and construction firm. Fluor has been the majority investor in the technology since 2011. Additionally, NuScale

---

45 Brigham, Lawson W, World's Most Capable Icebreakers: Russia's New Arktika Class, *Proceedings, U.S. Naval Institute*, May 2022, https://www.usni.org/magazines/proceedings/2022/may/worlds-most-capable-icebreakers-russias-new-arktika-class, accessed December 2024.

has received funding and support from the US Department of Energy, which has granted hundreds of millions of dollars to support the development and commercialisation of SMRs.

The basic NuScale design is a 77 MW module, which can be transported by truck. The design received certification from the US Nuclear Regulatory Commission in August 2020, making it the first SMR design to achieve this milestone. This was a significant step towards the first commercial deployment of this technology, a 462 MW plant planned for construction at the Idaho National Laboratory. Unfortunately, this project has been suspended due to cost overruns; however, a similar project in Romania is progressing as planned. Fluor completed Phase 1 front-end engineering and design (FEED) for the Doicesti facility in the fourth quarter of 2023 and commenced the Phase 2 FEED in July 2024. The US International Development Finance Corporation is currently considering US$1 billion of financing for the project, and assuming all goes to plan the project will be commissioned in 2029.[46]

The Rolls-Royce SMR is larger than conventional designs with an electrical output of 470 MW, but Rolls-Royce still categorises it as an SMR due to its modular design and manufacturing approach. Key components such as the reactor pressure vessel and steam generators will be prefabricated in factories allowing for the reactors to be produced more efficiently in a controlled environment, thus saving time and cost.

Rolls-Royce, with UK government support, plans to have its first SMR operational in the UK by the early 2030s. While specific locations have yet to be confirmed, Rolls-Royce has

46 NuScale Small Modular Reactor Fact Sheet, NuScale Power, LLC, 2021, https://nuscalepower.com, accessed 28 November 2024.

shortlisted several potential sites. These are generally on or near existing or decommissioned nuclear sites due to the existing infrastructure, regulatory framework and public support. Possible locations include Wylfa and Trawsfynydd in Wales and Sellafield in Cumbria, England.

Australia's plan to build a 235 GW all-renewables power system on its east coast is a significant undertaking, especially with a diverse mix that includes rooftop solar, utility-scale solar, onshore and offshore wind, flexible gas backup and hydroelectric power. This scale of renewable integration is among the largest planned globally, particularly as a standalone system.

One of the biggest challenges will be managing the intermittency of solar and wind resources. Ensuring grid stability without traditional baseload power sources requires advanced grid management technologies and backup. Denmark's and Ireland's power generation will rely heavily on renewables but as pointed out above they both have circa 120% firm power backup from neighbouring countries. Australia's east coast, with 10% firm power backup, will have nothing like that, so there will be shortages, particularly on days when the wind doesn't blow, and these shortages will become more frequent as coal-fired stations are phased out.

When this happens nuclear won't be an option because legislation currently prevents it. The other option will be more gas-fired power generation, in other words, more CO_2. In response to evolving global and regional security threats, Australians quite happily embraced the idea of nuclear submarines. American submariners don't glow in the dark and nor will Australian submariners. Now, in the face of an existential threat from global warming we should be equally happy to embrace nuclear as a carbon-free way of stiffening the backbone of our power supply system if it's warranted. But there can never be a serious discussion around nuclear while there is legislation preventing its use.

As coal-fired power stations are retired and firm power backup for wind and solar fades to 10%, the fragility of the renewables network will be exposed and the need for a Plan B will become obvious. More wind and solar will not be the answer. One hundred windmills produce no more electricity than one windmill when the wind isn't blowing. More gas-fired power stations won't be the answer either because we now know net zero is not good enough. Real zero is the only way to avoid ocean acidification.

On 10 October 2024 the Australian House of Representatives established a Select Committee on Nuclear Energy to consider nuclear power generation, including the deployment of SMRs in Australia. Part of its remit will be to examine the costs of deploying, operating and maintaining nuclear power stations. Table 10.1 demonstrates that the government's plan to transition to net zero with 'renewable energy connected by transmission, firmed with storage and backed up by gas-fired generation' could be replaced by a nuclear option at a lower financial cost, without gas-fired backup. It is not suggested that Australia should abandon its current plan for renewables entirely, but that nuclear becomes an important element of the mix. This will provide Australia with a robust power generation network, a path to real zero and, God forbid, an end to the Lucky Country's nightmare.

Made in the USA
Monee, IL
05 November 2026

67387146R00071